Food Microbiology and Food Safety

Practical Approaches

Series Editor:

Michael P. Doyle
Regents Professor of Food Microbiology (Retired)
Center for Food Safety
University of Georgia
Griffin, GA, USA

Food Microbiology and Food Safety Series

The Food Microbiology and Food Safety series is published in conjunction with the International Association for Food Protection, a non-profit association for food safety professionals. Dedicated to the life-long educational needs of its Members, IAFP provides an information network through its two scientific journals (Food Protection Trends and Journal of Food Protection), its educational Annual Meeting, international meetings and symposia, and interaction between food safety professionals.

More information about this series at http://www.springer.com/series/7131

Jennifer McEntire • Andrew W. Kennedy
Editors

Food Traceability

From Binders to Blockchain

Springer

Editors
Jennifer McEntire
United Fresh Produce Association
Washington, DC, USA

Andrew W. Kennedy
Institute of Food Technologists
Global Food Traceability Center
Chicago, IL, USA

Food Microbiology and Food Safety
ISBN 978-3-030-10900-4 ISBN 978-3-030-10902-8 (eBook)
https://doi.org/10.1007/978-3-030-10902-8

This Springer imprint is published by the registered company Springer Nature Switzerland AG
The registered company address is: Gewerbestrasse 11, 6330 Cham, Switzerland

Preface

Seldom, if ever, have government and consumers lauded the food industry's ability to trace food products. Instead, stories illustrating the time it takes to untangle a complicated supply chain during a foodborne illness outbreak abound, paired with studies evaluating food fraud and findings of recalled products still available on store shelves. At the same time, most consumers have at least one smartphone, tablet, laptop, etc. in a world where connectivity is a must-have. Why, then, is food traceability so challenging?

The development of this book began when food traceability received heightened attention due to the Food and Drug Administration (FDA) Food Safety Modernization Act (FSMA). However, new laws do not generally get people excited about changing their internal and external systems, enabling the sharing of data that they may not want shared. Instead, market drivers push change.

Big buyers have demanded improvements in food traceability before. It was buyers who pushed produce suppliers to implement the Produce Traceability Initiative (PTI). However, very few members of that "last mile" of the supply chain followed through on scanning cases, minimizing the benefit of the effort. Now, some buyers have vocally advocated for the use of Blockchain to share data. However, there are prerequisites to Blockchain, namely, that the appropriate data need to be collected, structured, and digitized.

With ever-changing consumer preferences, a tremendous diversity of food products is sourced from around the world. They may be mixed, sorted, extracted, and processed by local, and sometimes transient, operations, multinational corporations, and everything in between. Systems to capture, store, and share data need to accommodate the scale and sophistication of these operations while still being able to function as a system. Herein lies the utility of global standards. Most food traceability initiatives, such as PTI, the Seafood Traceability Initiative, and mpXML (for meat and poultry), leverage the GS1 system of standards. There are standards that govern identification of products, locations, and other data attributes, standards for bar codes and RFID tags that convey this information, and standards for electronic sharing of data. Global standards are necessary to address concerns that would otherwise exist at later points in the supply chain, when products bearing different types of identifiers are handled at distribution centers and beyond.

Although the focus of this book is traceability, it is important to note that traceability is a by-product of good recordkeeping. Despite some visible outbreaks, the food supply is quite safe, and all partners in the food supply chain work diligently to protect the public from foodborne illness. Promoting improved traceability from a food safety standpoint is therefore not effective, since in actuality very few companies will ever need to use their traceability systems to respond to a food safety concern. Instead, the benefits of good recordkeeping take many forms, depending on point in the supply chain. Efficiency and marketing benefits often provide a return on investment. Tangible benefits aside, public pressure has forced the issue of traceability past the tipping point, and the food industry will need to address the challenges and seize the opportunities described in this text.

This book would not have been possible without the willing contributions of the authors, as well as the support of Tejas Bhatt, previously with the Global Food Traceability Center and now with Walmart, and Jennie Stitzinger with the Institute of Food Technologists, who patiently formatted the chapters and pushed us to the finish line just as the race was really starting.

Washington, DC, USA Jennifer McEntire
Chicago, IL, USA Andrew W. Kennedy

Contents

Chapter 1
Introducing the Drivers and Complexities to Tracing Foods

Jennifer McEntire

Abstract Although there are several definitions of traceability, when they pertain to food, they generally describe information flowing backward and forward through the supply chain to understand product movement. Often this requirement relates to a food safety objective, but the information needed to follow a product through the supply chain is likely captured in several different places. Responsibility for generating, storing, and accessing traceability related information from farm to fork lies with many roles within a company, and many companies in a supply chain. The nature of supply chains, traceability objectives, and other nuances associated with food make it difficult to apply the approaches used to trace non-food products to the food sector. Although there are several drivers and enablers to improving traceability, currently each company needs to evaluate how these drivers motivate them to change their practices in ways that enhance visibility.

Keywords Outbreak · Traceback · Recall · Standards · Terminology · Definition

Traceability is the ability to trace products—or more accurately, the history and data related to product movement. The word is used in several contexts to mean several different things. In a laboratory, it may refer to the ability to verify that a piece of equipment is calibrated to an "official" standard. Sometimes it's employed to prove genetic lineage. In some arenas it's often linked with the "cold chain". In the context of this book, traceability and product tracing are terms used to describe the movement of food products throughout a supply chain, inclusive of growing, harvesting, repacking, manufacturing, repackaging, storage, distribution, and sale to the consumer. Regulations may not always require recordkeeping, or the same extent of recordkeeping, at each of these steps, but nevertheless, anywhere that a food product has come from or gone to is a contributor to the traceability record.

Even when limiting the scope of traceability to focus on food supply chains, there are still subcategories. Traceability can be described as "internal", "external"

J. McEntire (✉)
United Fresh Produce Association, Washington, DC, USA
e-mail: jmcentire@unitedfresh.org

J. McEntire, A. W. Kennedy (eds.), *Food Traceability*, Food Microbiology and Food Safety, https://doi.org/10.1007/978-3-030-10902-8_1

or "whole chain", each of which is described in subsequent chapters. It can be further focused on the trace forward or trace back direction.

Because traceability relates to the physical flow of products, it is often associated with supply chains and logistics. But when the food industry at large talks about traceability, it is often in the context of a food safety issue. For the public health community and regulators, food safety, especially during outbreak investigations, is a primary driver of improving traceability. While the food industry places food safety as a top priority, resources are generally allocated toward preventing contamination events from occurring; traceability is viewed as being reactive to food safety events. The food industry would rather avoid a food safety event in the first place, rather than relying on a traceability system after the fact to unravel the puzzle. Thus, the drivers to improve traceability within individual food facilities are different and focus on improved efficiency and other benefits as described in Chap. 4.

This chapter is intended to address the terms that are used and roles that relate to traceability. Many of the topics introduced in this chapter will be elaborated upon substantially in subsequent chapters. The chapter also provides a brief synopsis of why traceability matters, not just to regulators seeking to protect public health, but also to the food industry.

Traceability and Recalls

As will be exemplified throughout this book, traceability and recalls are not synonymous. However, in many settings the terms tend to be used interchangeably, and some food companies continue to base their confidence in their ability to trace product in the fact that they were able to quickly execute a mock recall. Real life shows us that performance in a mock recall is no indicator of the robustness of a traceability system. In part, this is because the term "traceability" encompasses both tracing backward in the supply chain (usually to find the source of a problem) as well as tracing forward (the typical path followed by a recall, to understand the distribution of potentially contaminated product). Both the trace back and trace forward processes are wrought with complexity and dysfunction, resulting in a sub-optimal traceability system in the United States that puts consumer health at risk, despite the best intentions of regulators and the food industry. Some of their differences are highlighted in Table 1.1.

Table 1.1 Comparison of trace back and trace forward aspects of traceability

Difference	Trace back	Trace forward
Objective	Identify product	Remove product
Directionality	Fork to farm	Farm to fork
Initiator (typically)	Regulator	Industry
Scope	Supply chain	1-forward customer
Timeframe for records	24 h	24 h

Trace Back

As indicated in Table 1.1 and explained in Chap. 3, the objective of a trace back investigation is to identify a common element in outbreak investigations. Regulators, working with data from public health officials, attempt to follow the pathways of food products suspected of causing foodborne illness, with the hope of finding a common product, or common ingredient, produced in a single location at a specific point in time. When multiple consumers in different locations become ill, and all report eating a relatively uncommon brand of a product, the trace back investigation rapidly focuses on the branded product. However, in more complex trace backs, consumers may have eaten multi-component products, of which none of the ingre dients bore a label (e.g., salad, salsa, lasagna, etc.). In those instances, each ingredient may need to be traced (again, from multiple starting points). Several chapters within this book highlight the difficulties encountered when trying to piece together data from one supply chain partner with another, and the traceability pilot studies conducted by the Institute of Food Technologists documented several of these challenges [4].

Trace Forward (Recall)

Companies often set aggressive targets for their traceability systems, such as identification of 99% of implicated product in 30 min, or 100% in 4 h. These target timeframes are substantially shorter than the 24 h window within which a company needs to provide records of the immediate previous supplier and subsequent recipient of a product that is afforded by FDA. Still, in 2009 it took over 4 months for manufacturers to identify the nearly 4000 food products containing peanut paste products from the Peanut Corporation of America that were potentially contaminated with *Salmonella.* More recently, spices, especially cumin, were identified as containing undeclared peanut allergens. Recalls related to cumin and cumin-containing products rolled out over the course of several months. Even if all 24 h passed before a facility shared who received potentially adulterated product, how could the identification of all outlets of these products take many months? There are several aspects of internal traceability and external traceability that impede the rapid and efficient tracing of food products, even after the problematic product has been identified.

Analysis and Consequences of Traceability Definitions

Although there are many definitions of traceability, when it comes to understanding the path followed by food and its constituents, the exact definitions convey the same general essence. Select definitions of traceability are provided below, and other reviews have summarized several definitions. For the means of this chapter, and for the book overall, the preferred definition is the one offered by Olsen and Borit (2013), who proposes the definition of traceability as: "The ability to access any or all information relating to that which is under consideration, throughout its entire life cycle, by means of recorded identifications" [5].

Codex Alimentarius is the international standards setting body. Standards are established through the deliberation of governments and are relied upon if there are disputes within the World Trade Organization. The Codex definition of traceability is synonymous with product tracing (similar to the interchangeable use of terms in this book). The terms refer to the ability to follow the movement of a food through specified stage(s) of production, processing and distribution. Further, Codex states that traceability is one tool "within a food inspection and certification system in order to contribute to the protection of consumers against food-borne hazards and deceptive marketing practices and the facilitation of trade on the basis of accurate product description" [3]. Thus, Codex identifies two drivers or uses of traceability. One pertains to food safety, and the other relates to authenticity. Codex states that as part of its design, "the traceability/product tracing tool should be able to identify at any specified stage of the food chain (from production to distribution) from where the food came (one step back) and to where the food went (one step forward), as appropriate to the objectives of the food inspection and certification system" [3]. Codex also acknowledges that traceability "does not in itself improve food safety outcomes" [3]. This is because traceability does not prevent food from becoming contaminated. However, rapid and reliable traceability systems can prevent the further distribution and consumption of contaminated food, which is beneficial to public health.

The US Food and Drug Administration (FDA) does not currently have a codified definition of traceability and in fact prefers to use the term "product tracing" in lieu of traceability. The US FDA Food Safety Modernization Act (FSMA) indicates that the objective of product tracing is to "prevent or mitigate a foodborne illness outbreak and to address credible threats of serious adverse health consequences or death to humans or animals as a result of such food being adulterated". FDA states that "a product tracing system involves documenting the production and distribution chain of products so that in the case of an outbreak or evidence of contaminated food, a product can be traced back to a common source or forward through distribution channels" [4]. Through regulation, tracing is accomplished if a facility registered with the FDA knows where a product came from, and to whom it was shipped. In the case of manufacturing, the regulation, based on the Bioterrorism Act of 2002, states that the input and output products should be able to be linked, although the wording affords flexibility in the rigidity with which this is accomplished. Regulatory requirements are discussed in Chap. 2, but it is worth noting that FDA, as a public health agency, clearly limits the view of traceability to one related exclusively to food safety issues. This contributes to the continued efforts of food safety professionals to improve traceability, even though the skill set typically required for food safety bears little relation to traceability.

The US Department of Agriculture (USDA) requires traceability of products under the jurisdiction of the department, although the traceability drivers change throughout the supply chain. The Animal and Plant Health Inspection Service states that "Animal disease traceability, or knowing where diseased and at-risk animals are, where they've been, and when, is very important to ensure a rapid response when animal disease events take place. An efficient and accurate animal disease traceability system helps reduce the number of animals involved in an investigation, reduces the time needed to respond, and decreases the cost to producers and the government" [7]. This definition, and the driver behind it, does not necessarily correspond to a food safety issue, but is still health related, by focusing on animal health. The approaches to limiting the spread of disease amongst an animal population and limiting the distribution of contaminated food have some similarities, but also some differences. For example, knowing that a diseased animal traveled through a certain location may raise concerns about other animals at that location, whereas the storage of a contaminated food product in one warehouse rarely implicates other products stored at the same location.

The USDA's Agricultural Marketing Service requires that food suppliers to the school lunch program also have traceability, called a Domestic Origin Trace. The requirement can be met either by following the supply chain pathway for each purchase, or by having a Domestic Origin Verification Audit Program showing that traceability systems are in place [8]. In this instance, authenticity is the objective of the traceability system; USDA uses the program to ensure that the food it procures was sourced in the United States. In this context, the definition of traceability which relates to product movement still stands, but the application is unrelated to food safety.

The USDA FSIS is the agency responsible for the safety of meat, poultry, and processed egg products. As described in Chap. 2, FSIS, through assorted rules, requires their regulated industry to keep transactional records, including bills of sale, invoices, bills of lading, and all receiving and shipping papers, which include identifying information such as contact information, weights, names, etc. Similarly to FDA, the use of the term, and objective of traceability, relates to protecting public health during a food related event.

While Codex Alimentarius sets global standards that are generally adhered to and relied upon by governments, there are other international standards that are well recognized and worthy of mention. Although ISO is a private standard, it is recognized by many governments around the globe, particularly in the European Union. The ISO 22005 definition of traceability states that it is the ability to follow the movement of a feed or food through specified stage(s) of production, processing, and distribution [2]. This definition does not address the reasons why one would want to follow the movement of food and feed products, nor does it suggest how the process might be accomplished.

The Global Food Safety Initiative (GFSI) publishes a Guidance Document to which audit schemes are benchmarked. The certification of a food facility to a GFSI-benchmarked audit scheme is considered by many to be the "gold standard" in food safety management. An element of the Guidance Document, and thus in each of the benchmarked audit schemes, is a traceability requirement [1]. The document states that, as it applies to food manufacturing (Sector E), "The standard shall require that the organization establish, implement and maintain appropriate procedures and systems to ensure:

- Identification of any outsourced production, inputs or services related to food safety,
- Product identification that includes, at a minimum, the name and address of the producer,
- A record of purchaser and delivery destination for all products supplied."

GFSI does not specify how products should be identified, nor the granularity of identification (e.g., production lot, shipment, etc.). There is also no required timeframe within which this information must be available. However, some of the audit schemes associated with GFSI do have more specific traceability requirements, and often have a requirement to demonstrate the effectiveness of the traceability system through mock recalls.

Traceability versus product tracing: The Codex Alimentarius definition uses both product tracing and traceability interchangeably, and this is the approach in this chapter and throughout the remainder of this book. Historically, the term traceability was preferred, but the United States, particularly the FDA, transitioned toward the term "product tracing" around 2009 to acknowledge that other applications of the term "traceability" pertained to following the lineage of an item, be it a laboratory standard, genetics of a seed stock, etc.

Responsibility for Traceability

As is evident in the definitions above, which in some instances convey the objective or reason for traceability, the responsibility for food traceability within a food production facility may vary. It can be challenging to identify a single individual within a facility who knows receiving, manufacturing/batching, and shipping, as well as has insight into food safety concerns that may trigger the need to track products. For this reason, both the initial architecture and establishment of a traceability system, as well as the process of obtaining information from it as needed, should be managed by a multidisciplinary team. The most common team members include the following:

- Food safety
  - As noted previously, traceability is in the spotlight when a food safety event occurs that necessitates the identification of the source of materials, or, later in an investigation, an understanding of where potentially contaminated materials were sold and how they were used. However, given the mechanics of traceability, the role of the food safety professional is peripheral and advisory to the rest of the team. The food safety professional should understand regulatory recordkeeping requirements and ensure that the system is designed with the degree of granularity and precision to fully protect public health while at the same time being able to pinpoint potentially affected product, limiting loss to the facility.

- Supply chain and logistics
  - Traceability is about product movement. Therefore, the role of the supply chain professional is key to good traceability.
- Information technology
  - Traceability is about product movement, as stated above. But that movement needs to be recorded in order for a traceability record to exist. While the food industry continues to use paper-based recordkeeping, electronic records are strongly preferred. Many companies use "legacy systems" that were custom built by IT professionals to house data, including data needed to establish traceability. As companies transition to other software solutions, IT professionals play a critical role in the selection and implementation of such systems.
- Accounting
  - Given that goods are purchased and sold, some traceability related information may be housed in accounting systems. In some situations, the party purchasing or selling the product may not be the same person who receives or ships the product. While "following the money" may sometimes be helpful, it is important to remember that traceability is about following the movement of physical objects.
- Procurement and sales
  - Those who purchase and sell food and feed products should have visibility into who the customers are and be able to determine the origin and destination locations of raw materials and finished products. If the product bears claims that need to be authenticated (such as organic, Fair Trade, etc.) procurement and sales staff have an added stake in traceability.
- Shipping/receiving
  - The individuals who receive and prepare bills of lading and handle physical product (including verifying accurate counts) are critical to ensuring accurate traceability. Bills of lading provide information about the actual "ship from" location, and need to be retained, and preferably maintained electronically.
- Operations
  - When establishing the traceability of a manufactured product (one that is comprised of multiple inputs, including situations where the product is repackaged), it is critical that the link between specific inputs and specific outputs be accurate. Operations is generally concerned with production efficiency and may use several of the internal systems used to calculate production efficiency. Raw materials or finished products that are unaccounted for introduce serious traceability challenges.

Food Traceability Compared to Other Sectors

Often when traceability challenges food safety investigations, questions are raised as to why parcel carriers can handle such volumes of packages, and customers can go online and enter a code to know precisely where the package currently resides. Automobile parts are stamped with traceability information. Why is tracking food so much harder?

There are several factors that make food traceability much more complex than tracing other types of products. Yet, this should not be used as an excuse to abandon all hope of improving the traceability of the overall food system. The accessibility of technology, combined with regulatory changes that may ultimately improve supply chain visibility, should enable more rapid and effective product tracing.

- Food supply complexity
  - Global
    - Like other industries, food traverses the globe; sometimes a few times; before ultimately being consumed. The global nature of the food supply in and of itself need not be a specific challenge to traceability. However, varying regulations pertaining to the information that should be associated with a product, inconsistencies in how this information should be formatted and communicated, and the variation in technological maturity around the globe contributes to the difficulties in tracing products worldwide. Of note, it is important to point out that some emerging economies have taken advantage of the availability of newer technologies to facilitate traceability, compared to some areas, such as the United States, that began developing during the Industrial Revolution and still have production plants that lack modern telecommunications capabilities.
  - Multiple handlers
    - Not only is the food supply global, but regardless of the physical distance food, and its components, travel, a finished food product may consist of numerous ingredients, some of which may have their own ingredients, and each of which may have been chopped, sorted, repacked etc. by several different supply chain members. Each entity along the circuitous path that is followed must capture traceability information and pass it along to the next supply chain member. One up-one down recordkeeping seems straightforward when there are only a few supply chain players but becomes much more difficult to handle when food and ingredients may have been transferred dozens of times.
  - No barrier to entry
    - Unlike many other industries that are viewed as having traceability systems superior to food (as described below), just about anyone can produce and sell food. Unfortunately, the result is that individuals can start a business in the food system without full awareness of the rules, regulations, and indus-

try standards that exist. Facilities that manufacture, process, pack or hold food regulated by FDA in the United States, or that send those products to the US, must register with the FDA. As of early 2016, FDA had received over 300,000 registrations [10]. There are over 6000 US-based meat, poultry and processed egg establishments inspected by USDA FSIS [9].

- Sheer volume
  - How many cars does a family of four typically purchase in a year? How many packages may be delivered to a specific street address? These figures pale in comparison to the number of food items eaten by a typical family in a year. The production of food is ongoing, and most food products enter and leave the supply chain, and consumer's homes, very quickly compared to other consumer products.
- Low profit margin
  - With a few exceptions, food producers generally derive profit because of the sheer volume of product sold, as indicated above. Compared to other products, food does not have a high margin of profit, making food producers resistant to innovate and invest in new technologies. However, the claim that the low profit margin in the food industry prevents the adoption of new technology is not entirely accurate. Rather, food producers need to see a return on investment, discussed in Chap. 4 to justify an initial outlay of capital.

McEntire et al. [4] profiled several industries and evaluated the state of traceability, as well as the traceability drivers, within those industries. While traceability within those industries has continued to evolve, the basic concepts discussed still stand. The publication raises an important point, which is that the motivation for traceability in other industries often differs from the motivation for food traceability. In some cases, such as for pharmaceuticals, verification of authenticity and the prevention of supply chain substitution has driven the traceability requirements. Other industries seek increased efficiency, and others may be responding to consumers' demand for increased information and visibility. As the food industry contemplates improved traceability, some sectors or companies may find that drivers in other industries apply to them, too [4]. As stated earlier, food safety is often a top-of-mind driver for traceability, but a broader evaluation of the benefits of improved recordkeeping should be fully explored so that business benefits are more readily derived.

Regulators Role in Traceability

Chapter 2 provides more details on the US regulations around food traceability. When public health is at risk, regulators swiftly attempt to identify the source of the problem. Regulators examine the entirety of the food supply chain, sometimes spanning the full spectrum from fork backward to the farm, to rule out suspect items and focus on those that seem to be common to multiple illnesses. Because of their perspective, they are often best positioned to identify thematic deficiencies in traceability. This is in contrast to individual members of the food industry who rarely have the opportunity to understand how the traceability information they maintain and share connects to a larger trace of the food supply.

As discussed in Chap. 2, there are regulations that dictate the types of records that must be maintained for various parts of the food industry, as well as timeframes for the submission of such information. The laws passed by Congress limit the scope of regulations that FDA and USDA FSIS can issue. However, just because an agency cannot force companies to submit relevant information more quickly, regulators—and the public—benefit from the voluntary cooperation during an outbreak or other investigation.

Traceability Drivers in the Food Industry

As eluded to in this chapter, there are several applications of traceability. One can expect to receive a wide variety of responses to the question "why is food product traceability important?" The background of the individual offering an answer will probably drive the response into one of the following categories:

- Traceability is needed for food safety
- Traceability helps improve operational efficiency
- The visibility that accompanies traceability allows us to communicate information customers are asking for
- Traceability helps authenticate claims

The types of benefits that can be realized by these and other applications of traceability are described in detail in Chap. 4.

Woven throughout this book are several high profile, and some less well-known, examples where poor traceability hindered outbreak investigations, exacerbating the public health outcome of food safety outbreaks.

The oft-cited example, described in great detail in Chap. 3, is the 2009 outbreak of salmonellosis in the United States that persisted in part due to the erroneous identification of tomatoes as the causative food vehicle, when it was later revealed that the source of contamination was likely peppers.

The Future of Traceability

There are several factors converging that increase the likelihood that food will continue to become more traceable.

First, recent advances in technology and infrastructure will enable better food traceability. For some companies, and in some parts of the world, the adoption and implementation may still be cost prohibitive. However, as with other applications of technology, we can expect that at some point in the future technology issues will cease to be a barrier to better traceability.

Second, many consumers want to know additional information about their purchases. There is increasing interest in the food that we eat, including where it comes from, the origin, belief attributes, allergens, ingredients, and other factors. Traceability is currently being used in some sectors as a marketing advantage. Several produce items now carry a two-dimensional bar code that can be scanned by consumers to reveal the date the item was harvested, location of harvest, and in some cases the farmer or crew that picked the item.

Third, the pressure to be lean and efficient continues to grow. While there have always been financial incentives to a more efficient operation, we are now more cognizant of the scarcity of resources. Successfully feeding nearly ten billion people by 2050 will require that food production be as efficient as possible, and traceability provides the visibility to identify opportunities to improve efficiency [6].

Last, for industry members who are not incentivized by marketing opportunities or gains in efficiency, the "big stick" of regulation may ultimately force improvements in traceability upon the food industry. However, as explained in Chap. 2, the limitations of the laws currently in place in the United States make it unlikely to achieve a highly functional, effective traceability system.

Various food sectors have conceptualized, and in some cases begun to implement, changes in the nature, capture, and exchange, of traceability related information. The benefits of traceability, both for individual companies as well as society at large, will likely outweigh the immediate costs.

References

1. Global Food Safety Initiative [Internet]. Issy-les-Moulineaux: GFSI guidance document; c2013 [cited 2016 Jun 14]. Available from: http://www.mygfsi.com/images/mygfsi/gfsifiles/gfsi_guidance/GFSI_Guidance_Document.pdf
2. ISO [Internet]. Geneva: ISO 22005:2007; c2011 [cited 2016 Jun 14]. Available from: http://www.iso.org/iso/catalogue_detail?csnumber=36297
3. Joint FAO/WHO Codex Alimentarius Commission, Food and Agriculture Organization of the United Nations, Joint FAO/WHO Food Standards Programme [Internet]. Rome; c2006 [cited 2016 Jun 14]. Available from http://www.codexalimentarius.net/input/download/standards/10603/CXG_060e.pdf
4. McEntire J, Arens S, Bernstein M, Bugsu B, Busta F et al Product tracing in food systems: an IFT report submitted to the FDA, volume 1: technical aspects and recommendations. Compr Rev Food Sci Food Saf [Internet]. 2010 Jan [cited 2016 Jun 14]. Available from: http://onlinelibrary.wiley.com/doi/10.1111/j.1541-4337.2009.00097.x/abstract http://www.fda.gov/downloads/Food/GuidanceRegulation/UCM341810.pdf
5. Olsen P, Borit M. How to define traceability. Trends Food Sci Technol [internet]. Feb 2013 [cited 2016 Jun 16]; 29(2):[about 8 pgs]. Available from: http://www.nofima.no/filearchive/tifs1412.pdf

6. United Nations [Internet]. New York: World population projected to reach 9.6 billion by 2050; c2013 [cited 2016 Jun 14]. Available from: https://www.un.org/en/development/desa/news/population/un-report-world-population-projected-to-reach-9-6-billion-by-2050.html
7. United States Department of Agriculture [Internet]. Washington: Animal disease traceability; c2016 [cited 2016 Jun 14]. Available from: https://www.aphis.usda.gov/aphis/ourfocus/animalhealth/sa_traceability/!ut/p/z1/rVJdb4IwFP01eyS3BYT6CIZMF90S-FZW-kGsH0g0LYsX571d9lPixZH1p7s2595yeU-CwAq6wlRvUslJYmjrhXjr-cAc-0nBFzRyEJoqk_iaKRTaY2LK8Ai6FnAPN4_s5oOOrbwJ-ZJzdOQB7NL4ADF0rX-uoAE60LuU1EpnSmdlnLdYHN6IXtMq0OT5pU47C8VKrnFMi0yLHVx6egGRY-ZrWUp9Om-shfyExBaUsbwnLEJ9tFyP2hYyz7HWPsv6jDDW7_kdBzoS-f0HLs-98dzZQpwPomvyIJDEi_VsUr9SFZSuzI8SqaowzMPujB0MCb4-SMl9Jfu12PDB5nRP60bD6p8DMaruZDCYbIxp1YUmVV7C6RtXbON4y52R9T9lxnhdlOw5-AUi28VE!/dz/d5/L2dBISEvZ0FBIS9nQSEh/?urile=wcm%3Apath%3A%2FAPHIS_Content_Library%2FSA_Our_Focus%2FSA_Animal_Health%2FSA_Traceability
8. United States Department of Agriculture [Internet]. Washington: Domestic origin verification audit program (DOV); c2013 [cited 2016 Jun 14]. Available from: https://www.ams.usda.gov/sites/default/files/media/DOV%20presentation.pdf
9. United States Department of Agriculture [Internet]. Washington: Meat, poultry and egg product inspection directory; c2016 [cited 2016 Jun 14]. Available from: http://www.fsis.usda.gov/wps/portal/fsis/topics/inspection/mpi-directory
10. U.S. Food and Drug Administration [Internet]. Silver Spring: Registration statistics; c2016 [cited 2018 July 28]. Available from: http://www.fda.gov/Food/GuidanceRegulation/FoodFacilityRegistration/ucm236512.htm

Chapter 2
Tracing the Food Safety Laws and Regulations Governing Traceability: A Brief History of Food Safety and Traceability Regulation

Shawn K. Stevens

Abstract Laws and regulations related to food safety, including traceability, have evolved substantially over the last 100 years. Although both the USDA FSIS- and FDA-regulated food industries are already subject to traceability requirements, the agencies still struggle to rapidly and effectively conduct traceback investigations. As a result, the agencies are pressing forward to strengthen the existing rules, regulations, and policies governing traceability. On the USDA FSIS side of the fence, the agency has enacted more stringent recordkeeping requirements for beef ground at retail and is beginning traceback investigations into the original source of pathogens when they are discovered in ground beef products more rapidly and with less data than ever before. On the FDA side of the fence, the agency has been authorized by Congress to create enhanced recordkeeping requirements to better trace high risk foods. Moving forward, it remains to be seen whether these new initiatives will in fact enhance the overall traceability of food products. Although in some ways the new requirements may result in some improvements, significant challenges may continue to persist. In this regard, to the extent any hindrances to traceability ultimately continue, they may be related less to the rules on the books, as opposed to the inherent difficulties in enforcing them across an extremely diverse industry, and a persistent lack of any globally harmonized requirements.

Keywords Law · Regulation · Liability · Bioterrorism Act · Federal meat inspection Act · FDA · USDA · Regulatory requirements · Recordkeeping · Food safety modernization Act

---

S. K. Stevens (✉)
Food Industry Counsel, LLC, Milwaukee, WI, USA
e-mail: stevens@foodindustrycounsel.com

J. McEntire, A. W. Kennedy (eds.), *Food Traceability*, Food Microbiology and Food Safety, https://doi.org/10.1007/978-3-030-10902-8_2

Introduction

The laws and regulations which govern the traceability of the food we enjoy have a very long lineage. Many of the federal laws and regulations that govern the production, warehousing and distribution of food in the United States were first enacted over 100 years ago. In turn, for more than a century, those standards have evolved and matured as lawmakers and regulators have strived to protect American consumers by continuously (and, in some cases, tirelessly) addressing emerging and diverse food safety-related public health challenges.

The traceability of the food that makes its way onto our dining room tables has always been of critical interest to American consumers. Although, in many ways, the degree of interest has evolved and changed over the years, what has remained constant is the public's concern over the ultimate and underlying quality and safety of its food.

Indeed, the journey experienced by the food industry and consumers over the last 100 years, as it relates to food safety and quality, was not easy or without controversy. At the dawn of the last Century, through advances in science and technology, the food industry, for the first time, began to understand the processes through which food could be made more accessible to the public. With the emergence of rapid transnational shipping, improved preservation techniques and the ability of media and, by extension, advertising and marketing campaigns, to reach an increasing number of consumers, the food industry quickly revolutionized. For the first time in the history of the nation, food processors could viably and quickly ship for display and sale anywhere in the nation a broadening array of perishable products.

While the interstate shipment of food products began to grow, however, the rules and regulations governing the underlying safety of those products were woefully inadequate. In the absence of a unified federal approach to food safety regulation (there had not yet been a perceived need), America's food laws were implemented primarily at the state and local level. As can be expected within a growing nation, however, an increasing number of industrial advances soon began to quickly outpace the limited state and local regulations. In the meat industry, in particular, laws defining what constituted "adulteration" or "misbranding" were determined, if at all, by each individual state. Moreover, what was forbidden inside the borders of one state was in many cases entirely lawful in another.

This hodgepodge of inconsistent regulation soon inspired concerned citizens, consumer groups, and social reformers to voice their concerns. Without a national approach to food safety (and a single set of rules), American citizens in the various states had no confidence in the origins or safety of the food they were eating.

And, with this backdrop, numerous additional technological advances outside of the food industry became instrumental in inspiring incredible change as well. The emergence of inexpensive newspapers being distributed across the nation gave both individual consumers and organized consumer groups greater opportunity to voice their concerns about food safety, food quality, and related issues. In turn, social reformers, who otherwise would have remained unheard, suddenly found themselves able to reach a much broader audience. The most famous example, of course, was Upton Sinclair.

In 1906, when Sinclair's book was published, meat producers in the United States were virtually unregulated. As a result, American consumers had no real understanding about the origins or safety of their meat products. In his novel, however, Sinclair described in punishing detail the unsanitary conditions then prevalent in large slaughter plants. The book described unwholesome and diseased carcasses being processed for use in food, food production taking place in unhealthy and insanitary conditions, food and meat products being made from undisclosed mixtures of animals and animal parts, processing taking place in rodent- and insect-infested conditions and meat and other food products coming into contact with dirt, mud, feces, poisons, and other contaminants present in the packing plants – prior to being distributed for sale.

Learning about such conditions outraged American consumers, and the resulting demands for change from the public soon became too loud for Congress to ignore. Inspired by a broadening national chorus insisting on better food safety and quality standards, the federal government realized that a uniform food safety policy (*i.e.*, a single set of rules) was essential to protect both the national economy and the health of American consumers.

Federal Laws and Acts Relating to Food Traceability

The Federal Meat Inspection Act

In 1906, Congress responded to the public uproar by passing the 1906 Wholesome Meat Act. The 1906 Wholesome Meat Act (now known as the Federal Meat Inspection Act, 21 U.S.C. § 601, et seq.) [16]. was designed to provide assurance to the public that its food would be produced under sanitary conditions, and to more effectively suppress and prevent the spread of diseased livestock and contaminated food. With the sweep of a pen, the federal legislation for the first time ever required continuous federal inspection within meat packing plants to ensure that all meat products sold in interstate commerce would be: (1) produced under sanitary conditions; (2) not adulterated; and (3) properly labeled. In addition, by mandating that all meat products sold in interstate commerce be labelled with the federal seal of inspection, the 1906 Wholesome Meat Act could in many ways be said to represent the birth of food product traceability.

Today, the Federal Meat Inspection Act ("FMIA") is enforced by the Federal Food Safety and Inspection Service ("FSIS"), an agency falling within the jurisdiction of the United States Department of Agriculture ("USDA"). In meat packing plants, FMIA continues to require continuous, on-site inspection of the entire slaughter and processing process. In addition to its inspection responsibilities, FSIS also closely regulates product formulation and labeling. And, over the decades,

FSIS's authority has expanded further to oversee harvest and processing under the Poultry Products Inspection Act ("PPIA") and egg production under the Egg Products Inspection Act ("EPIA") [1, 12].

In the 1990s, in order to provide additional incentives for regulated establishments to implement broader measures to prevent the bacterial contamination of their products, FSIS adopted a new system of inspection to protect against microbiological hazards in raw animal foods. The agency felt that a new system was needed, in part, because the then-existing system of organoleptic inspection, which was based upon sight, touch and smell, "did not directly target the reduction of pathogens" [11]. Unlike the organoleptic inspection scheme, the Hazard Analysis Critical Control Point ("HACCP") methodology was designed to build pathogen prevention directly into the inspection program.

The HACCP concept was first developed by Dr. Howard Baumann, who worked for Pillsbury, as a quality control program governing the processing of food for the United States space program [8]. Under the HACCP methodology, a food producer would identify all hazards reasonably likely to affect the safety of a food product during production (including the introduction of chemical, physical and biological contamination), and then establish "critical control points" where specific interventions could be used to prevent, reduce or eliminate any identified hazards. By the mid-1990s, USDA recognized that HACCP could provide a framework for meat harvest and processing facilities to develop and implement controls to reduce emerging food safety hazards such as dangerous pathogens. The Final Rule adopting HACCP was introduced on July 25, 1996 [10].

To this day, the USDA maintains a physical presence within meat processing facilities and continues to closely enforce the HACCP regulations. This system has proven quite effective in reducing the incidence of pathogens in raw animal products. In 2007, as industry was still learning how to best apply HACCP within meat facilities to control *E. coli* O157:H7, for instance, the meat industry experienced a total of 22 recalls relating to the presence of *E. coli* O157:H7 in an astonishing 35,000,000 pounds of beef. Only 5 years later, in 2012, the industry had improved so much using the HACCP methodology that it experienced a total of only four recalls (for *E. coli*) involving a mere 25,000 pounds of beef.

Ensuring the protection of the public health, however, requires more than just overseeing the production of safe food. This is because, despite best efforts, pathogens remain elusive and can still contaminate food products. If contaminated raw animal products are not adequately handled and prepared by consumers, then those consumers can become sick. If ready-to-eat products are contaminated post-lethality treatment, then they can also cause outbreaks. As discussed in the previous chapter, when outbreaks occur, the government's ability to identify the common food source causing illnesses has become quite robust. For this reason, once an offending product has been identified, the government can only protect the public if it has the ability to quickly identify the producer (or packer) that was the original source of the contamination, identify and contain any affected lots, urge the producer to issue a voluntary recall, and then (at the same time) provide a notice to the public regarding the recall and any implicated products.

For this reason, many of the current laws and regulations governing the safety of our meat supply already boast significant traceability elements. The statutory laws and regulations adopted pursuant to FMIA, PPIC and EPIA, for instance, require official establishments and retailers to "keep records that will fully and correctly disclose all transactions in their business subject to the [relevant] Act[s]" [14, 15]. Thus, any time any livestock or food products derived from livestock are transferred between two parties for sale, these standards require those establishments and retailers to maintain traceability records which include, but are not necessarily limited to, bills of sale, invoices, bills of lading, and all receiving and shipping papers.

With respect to each transaction, these records must provide information which, at the very least, identifies the name or description of the article that is the subject of the transaction, the net weight of the article, the number of outside containers, the name and address of the buyer and seller, and the date and method of shipment. As noted, in addition to governing the sale of articles, these requirements also apply with equal force to any transactions that involve animals or livestock. Although there are always at least some exceptions, as a general matter, federal establishments must maintain copies of these records for a period of 2 years after December 31 of the year in which the transaction to which the records relate has occurred [13]. In turn, these requirements give FSIS the basic tools to trace the distribution of meat and poultry products through commerce, as they pass through the chain from harvest to retail.

To enhance product integrity and tracking, establishments are also required by FSIS to maintain records relating to packaging, labeling, food product formulations and daily production [19]. In addition to these documents, establishments must also maintain records relating to the execution and monitoring of their standard sanitation operating procedures and their HACCP programs, including information relating to production, production lots and pre-shipment reviews [14, 18].

In turn, 9 C.F.R. Part 325 prescribes additional records requirements for the transportation of meat products [20]. These regulations state, in relevant part, that meat products cannot be shipped in commerce unless they are properly labeled and carry the federal seal of inspection. Although there are certain exceptions, the processing records relating to refrigerated products must be maintained for a year. If the products are frozen or considered shelf-stable, however, the records must be maintained by the establishment for 2 years. By virtue of its regulatory and enforcement authority, FSIS has the ability to request and be given access to these records at any time.

The labeling requirements for meat products also facilitate traceability in many important ways. The eight basic requirements for product labeling include: (1) the product name; (2) a copy of the federal inspection legend and establishment number; (3) a handling statement, if needed (relating to the need for refrigeration, for instance); (4) a net weight statement; (5) an ingredients statement; (6) an address line; (7) nutrition facts, and (8) the federally-mandated safe handling instructions. All federal establishments are required to register with the government, and every product label must carry the federal establishment number assigned by FSIS to that facility. As a result of these requirements, in the event of an outbreak or positive product sample, FSIS can in most instances quickly identify the exact location where the product was processed.

Not all traceability systems, however, are perfect. Despite these relatively comprehensive recordkeeping and labeling requirements, FSIS has been impeded in many cases in its efforts to effectively trace ground beef products that have been involved in an outbreak back to their original source. Namely, this is because although retailers (like federally-regulated establishments that harvest and process meat products) are required to maintain records relating to any "transactions" involving meat products, they have not historically been required to maintain records that pertain to the further processing of the products at retail. Thus, although many retailers such as grocers would further grind and then repackage the ground beef they receive from an establishment, or would grind and repackage internal bench trimmings, they were not required to identify with particularity the specific source material used in any specific lot. Without such records, FSIS could not always conduct timely and effective traceback investigations when an outbreak was occurring.

In an effort to help address these concerns, the agency published a federal register notice in 2002 which outlined the information that the agency considered important for effective traceback and trace-forward activities [2]. In the notice, FSIS stated that if the agency was made aware of any raw ground beef product sold at retail that was tested and confirmed to be positive for *E. coli* O157:H7, it would begin collecting from the retailer the names and numbers of the establishments supplying the source materials for the ground beef sampled, as well as the supplier lot numbers and production dates. After initiating these efforts, however, FSIS discovered that many retailers still refused to maintain adequate records which would enable the agency to identify the source materials of the batches of product being processed.

Thus, in 2009, FSIS took additional steps by providing guidance to the retail industry which requested that retailers take affirmative steps to implement and adopt a practice of recordkeeping that would allow FSIS to better trace these products. In particular, the agency recommended that retail stores keep records of the lot or batch number of the source materials used by the retailer to prepare any raw ground beef, as well as the name and type of product produced, the manufacturer of the source material, the source material product code or pack date, and the supplier's establishment number.

Despite these efforts, in the years that followed, the Agency continued to experience significant difficulty in tracing contaminated products sold at retail back to the original processor or harvest facility. In nearly 30 separate foodborne illness outbreak investigations led by FSIS dating back to 2012, the agency found that in nearly a dozen of the outbreak investigations, there were no retail grind records available to assist with the ongoing traceback efforts, and that although some records were available in six of the investigations, the records proved to be inaccurate or incomplete.

As a result, FSIS concluded in 2014 that voluntary record keeping by retail facilities had proven to be ineffective. FSIS efforts were further frustrated because, in some cases, official establishments who were processing ground beef were also not maintaining adequate grind records identifying the suppliers of raw materials used in each individual batch. As a result, FSIS proposed to amend 9 C.F.R. Part 320 in July of 2014 to require retailers and federal establishments to maintain more detailed traceability records. On December 14, 2015, after considering substantial public comment, FSIS published the final rule "Records to be Kept by Official Establishment and Retail Stores That Grind Raw Beef Products." The final rule requires all retail stores and official establishments that grind raw beef products to keep records which identify with specificity the supplier of all source materials that they used in the preparation of each lot of ground beef.

In particular, the new rule requires retailers and establishments to document: (1) the establishment numbers of the establishments supplying the materials used to prepare each lot of raw ground beef product; (2) all supplier lot numbers and production dates; (3) the names of the supplied materials, including beef components and any materials carried over from one production lot to the next; (4) the date and time each batch of ground beef product is produced; and (5) the date and time when any grinding equipment or other related food-contact surfaces are cleaned and sanitized [17]. These mandates have significantly improved FSIS' ability to trace product from the consumer level back to the original source.

In addition to these initiatives, FSIS has also changed the way it conducts its traceback investigations. In the past, if ground beef processed at a federal establishment or at retail tested positive for an adulterant, FSIS would notify the suppliers of the raw beef trim used to process the ground beef in question. Other than sending an email notification, however, FSIS rarely conducted any additional follow-up to determine the original cause of the contamination.

Under FSIS' new policy, the agency is now immediately sending Enforcement Investigations and Analysis Officers ("EIAOs") to the supplying establishments in question. The EIAOs are now required to perform an investigation at the facility in question to identify the likely cause of contamination. EIAOs will accomplish this task by scrutinizing the establishment's operational and microbiological testing records to find any evidence of insanitary conditions or product contamination. If the EIAOs determine that the facility is experiencing a broader food safety issue, they may ultimately conclude that a recall of raw beef trim (or any affected beef primals) is warranted.

As detailed above, FSIS has a wide range of regulatory tools available to facilitate the traceability of meat, poultry and other regulated food products. Nevertheless, moving forward, we anticipate that FSIS will continue to aggressively pursue additional regulatory policy initiatives that are designed at their core to make the traceability of the FSIS-regulated food supply, at all levels, as effective and seamless as possible.

The Federal Food, Drug and Cosmetic Act

Like the FMIA, the original Pure Food and Drug Act (now known as Federal Food, Drug and Cosmetic Act) was also enacted in response to the public's concerns about food safety and quality and the resulting broadening social pressure [6]. While the FMIA (which began as the 1906 Wholesome Meat Act) was designed to oversee the production and distribution of meat, the 1906 Pure Food and Drug Act was created to increase confidence in many other foods. Here too, Congress was driven to action following several well-publicized scandals involving the food industry, including one where American soldiers serving in the Spanish-American war were allegedly sickened by embalmed foods. Additionally, with advances in chemistry, it was becoming more common for the food industry to use various additives and chemicals to preserve foods. These activities, of course, also triggered additional debates involving the merits of substitute foods, such as margarine for butter, and the use of questionable "ingredients" such as coal tar, borax, and various food colors.

Thus, in 1906, the Pure Food and Drug Act, which was a companion Act to the Wholesome Meat Act, was passed by Congress and implemented by the Bureau of Chemistry, a division of USDA [5]. Although this Act did not require full-time physical "inspection" within food processing facilities, it did provide standards for the safety of all food products other than meat and poultry (which are regulated by FSIS under the FMIA). As envisioned by Congress, the Pure Food and Drug Act delegated authority to the Secretary of Agriculture to develop specific standards of adulteration. In response, the Pure Food and Drugs Act, as originally enacted in 1906, forbade adulteration and misbranding of foods in interstate commerce. This Act was amended several times before being replaced in 1938 by the more stringent Federal Food, Drug and Cosmetic Act ("FDCA") [6]. Overall, the new regulations developed over the years tracked closely those laws and regulations governing the United States meat supply.

Today, FDCA is enforced by the Federal Food and Drug Administration ("FDA"), an agency falling within the jurisdiction of the Department of Health and Human Services. Currently, FDA has jurisdiction over approximately 80% of the domestic and imported foods sold in interstate commerce, and seeks to ensure that such products are safe, nutritious, wholesome, and adequately labeled. In turn, the adulteration standards in FMIA and FDCA are virtually identical. Thus, what constitutes adulteration under one act will in most instances also constitute adulteration under the other. Although FDA, unlike FSIS, does not require or mandate the existence of continuous inspection in food processing facilities, it has jurisdiction (and conducts periodic inspections) where food is manufactured, processed, packaged, and held.

In early 2011, Congress substantially broadened FDA's power and authority to govern the production and distribution of food when it passed the Federal Food Safety Modernization Act ("FSMA"). [7] Recognizing how HACCP significantly improved the overall safety of our meat supply, as detailed above, FDA successfully petitioned Congress to extend meat-industry-style HACCP to all other segments of the food industry. Prior to FSMA, the only FDA-regulated products produced under mandatory HACCP systems were juice and seafood, and arguably low acid canned foods. With the passage of FSMA, FDA requires producers of *all* types of food products regulated by FDA to develop Hazard Analysis and Risk-based Preventative Control ("HARPC") plans, and then closely follow those written plans within their facilities.

The types of records that FDA will ultimately require companies to keep enhancing traceability, however, still remain somewhat vague. This is because, although certain standards already exist with respect to recordkeeping requirements and the agency's ability to access a company's records, FDA has yet to articulate with particularity the specific types of additional traceability records that it will expect food processors to maintain, especially as it relates to high-risk foods. We anticipate that the agency will provide additional guidance and propose additional rules relating to the agency's recordkeeping expectations as it works to finalize the remainder of the proposed regulations that have already been published.

Although we are still waiting on specifics, we do know that, with respect to food products or ingredients, the agency does currently require food processors and other supply chain members to maintain records that would enable FDA to quickly identify the immediate previous source of any food product or ingredient, and the immediate subsequent recipient. This standard was first adopted in 2002 as part of the Public Health Security and Bioterrorism Preparedness and Response Act and provides the current basic recordkeeping requirement as it relates to the traceability of food.

The FDA's recordkeeping regulations which were subsequently promulgated detail with specificity the types of records that must be maintained by companies that both process and transport food products. Pursuant to the regulations, nontransporters, which are companies that own, hold, manufacture, process, pack, import, receive, or distribute food for purposes other than transportation, must maintain records relating to both the immediate previous sources and the immediate subsequent recipients of any food [22]. In turn, under the regulations, transporters are defined as persons who have possession, custody, or control of an article of food in the United States for the sole purpose of transporting the food, whether by road, rail, water, or air. Transporters also include any foreign person that transports food in the United States, regardless of whether that foreign person has possession, custody, or control of that food for the sole purpose of transporting that food.

In order for a nontransporter to satisfy the recordkeeping standards relating to "the immediate previous source" of a food product, a nontransporter must establish and maintain the following information: (1) the name of the firm, address, telephone number and, if available, the fax number and e-mail address of the nontransporter's immediate previous source, whether domestic or foreign; (2) an adequate description of the type of food received, to include the brand name and specific variety (e.g., brand x cheddar cheese, not just cheese; or romaine lettuce, not just lettuce); (3) the date the nontransporter received the food; (4) for persons who manufacture, process, or pack food, the lot or code number or other identifier of the food (to the extent this information exists); (5) the quantity and how the food is packaged (e.g., 6 count bunches, 25 pound (lb) carton, 12 ounce (oz) bottle, 100 gallon (gal) tank); and (6) the name of the firm, address, telephone number, and, if available, the fax number and e-mail address of the transporter immediate previous source (the transporter who transported the food to the nontransporter) [23].

In turn, in order for a nontransporter to satisfy the recordkeeping standards relating to "the immediate subsequent recipients" of a food product, a nontransporter must establish and maintain the following information: (1) the name of the firm, address, telephone number, and, if available, the fax number and e-mail address of the nontransporter immediate subsequent recipient, whether domestic or foreign; (2) an adequate description of the type of food released, to include brand name and specific variety (e.g., brand x cheddar cheese, not just cheese; or romaine lettuce, not just lettuce); (3) the date the non-transporter released the food; (4) for persons who manufacture, process, or pack food, the lot or code number or other identifier of the food (to the extent this information exists); (5) the quantity and how the food is packaged (e.g., 6 count bunches, 25 lb. carton, 12 oz. bottle, 100 gal. tank); (6) the name of the firm, address, telephone number, and, if available, the fax number and e-mail address of the transporter immediate subsequent recipient (the transporter who transported the food from the nontransporter); and (6) the records must include information reasonably available to the non-transporter to identify the specific source of each ingredient used to make every lot of finished product [24].

Transporters are also required to maintain certain records which disclose the origin and destination of the foods they carry [25]. Although there are many different alternatives in the rules which transporters of food products can follow in order to comply with the existing laws, each transporter in essence must have information that will include the identity of the previous source of the food and subsequent recipient, the origin and destination points, a description of the food products transported, the route of movement, and the any transfer points through which the product may have passed. Ultimately, these requirements provide assurance that the core records which are essential to any traceability effort are both created and maintained.

To enhance traceability further, however, FDA also needs the authority to access any records that are created and maintained. For this reason, when the Public Health Security and Bioterrorism Preparedness and Response Act (commonly known as the Bioterrorism Act) was enacted in 2002, the legislation also added section 414 to the FDCA. This new section 414 gave FDA expansive new authority to access a company's records. Under the new law, the FDA is now permitted to have access to any records relating to any food product that the agency reasonably believes is adulterated and which would present a threat of serious adverse health consequences or death to humans or animals. In turn, when FSMA was enacted 9 years later in 2011, the new legislation gave FDA increased authority, permitting the agency to access additional records beyond only those records relating to the specific suspect article of food in question, to any records relating to any other article of food that the FDA reasonably believes is also likely to be affected in a similar manner.

More recently, on April 4, 2014, FDA published a final rule which details the authority FDA believed was delegated to the agency under FSMA [3]. In the final rule, FDA articulates the following powers as they relate to the agency's access to records:

> [I]f FDA believes that there is a reasonable probability that the use of or exposure to an article of food, and any other article of food that FDA reasonably believes is likely to be affected in a similar manner, will cause serious adverse health consequences or death to humans or animals, ... [then] FDA shall have access to the records that are needed to assist FDA in determining whether the food is adulterated and presents a threat of serious adverse health consequences or death to humans or animals, ... or whether there is a reasonable probability the use or exposure to the food will cause serious adverse health consequences or death to humans or animals.

Thus, under the FDA's new record access rules, FDA has been granted (and in turn has refined for itself) extremely expansive new powers which entitle the agency to access and review any records it deems necessary to not only effectively trace a product to its source, but also to determine whether that product was in fact handled or processed in such a way as to create a risk to public health.

Under this standard, FDA has consistently interpreted its powers to access records very broadly. In a guidance document published the same day as the new rule, FDA stated that its authority to access a company's records extends to "any records relating to the "manufacture, processing, packing, transporting, distribution, receipt, holding or importation of any food believed to be affected and any other article of food believed to be affected in a similar manner" [21]. In these guidance materials, FDA stated further that these records include, but are not necessarily limited to, the following: (1) manufacturing records; (2) raw material (ingredients and packaging) receipt records; (3) product distribution records; (4) product inventory records; (5) test records; (6) recall records; (7) reportable food records; (8) customer distribution lists; and (9) complaint and adverse event records.

In addition to modifying the recordkeeping and record access standards under the existing laws, FSMA also gave FDA new powers to explore and create additional traceability requirements; in particular, the legislation created a new section 204, titled "Enhancing Tracking and Tracing of Food and Recordkeeping". Section 204 has two major components to enhance food product traceability. First, the law mandated that FDA in coordination with other agencies and the food industry commission and complete pilot projects in order to explore and evaluate the most appropriate methods and technologies for rapid and effective tracking of foods. Second, the legislation required FDA to publish a notice of proposed rulemaking designed to establish new recordkeeping requirements for high risk foods designed to enhance the traceability of those products.

On September 7, 2011, FDA announced that that it had commissioned the Institute of Food Technologists (IFT) to conduct two pilot projects involving foods (and ingredients) that had been implicated in previous outbreaks. The projects were designed to explore various methods for rapid and effective tracking and tracing of those food products, the types of data that would be useful for tracking foods, the ways to connect various points in the supply chain, and how to best make the resulting data available to FDA. While conducting these pilots, IFT consulted with the food industry, USDA, and various state agencies and consumer groups, on the various challenges and solutions relating to food traceability. IFT issued its final report in August 2012 [9].

The final report contained IFT's recommendations to FDA on how best to improve the tracking and tracing of food. In the report, IFT recommended that rather than creating enhanced recordkeeping requirements for high risk foods only, that FDA should instead "establish a uniform set of recordkeeping requirements for all FDA regulated foods and not permit exemptions to recordkeeping based on risk classification" [9]. The report also recommended that FDA require any firms that manufacture, process, pack, transport, receive, hold or import food to identify and maintain records identifying critical track and trace information, including key data elements (KDEs) and critical tracking events (CTEs), and to develop, document and exercise a product tracing plan.

Two years later, in February 2014, FDA published a draft report outlining a proposed methodology that would enable an agency to identify high risk foods [4]. The methodology proposed by FDA for identifying high risk foods was a score-based system designed to enable FDA to make high risk determinations based upon the historic public health significance of the food in question with respect to confirmed outbreaks and foodborne illness cases, as well as several influencing food and food processing related variables and factors. Using this methodology, the FDA would be specifically required to consider: (1) the known safety risks of a particular food, including the history and severity of foodborne illness outbreaks attributed that such food, taking into consideration foodborne illness data collected by the Centers for Disease Control and Prevention; (2) the likelihood that a particular food has had a high potential risk for microbiological or chemical contamination, or would support the growth of pathogenic microorganisms due to the nature of the food or the processes used to produce the food; (3) the point in the manufacturing process of the food where any contamination is most likely to occur; (4) the likelihood of contamination and steps taken during the manufacturing process to reduce the possibility of contamination; (5) the likelihood that consuming a particular food would result in a foodborne illness; and (6) the likely or known severity, including health and economic impacts, of a foodborne illness attributed to a particular food [4]. Although FDA invited public comment on the draft methodology, the agency has not yet published any amendments or proposed rules relating to the methodology itself, or any formal specific high-risk determinations.

In addition to creating mandates relating to the identification of high risk foods, as detailed above, section 204 also required FDA to publish rules mandating additional record keeping requirements for high risk foods. Although FDA has not yet published the requisite proposed rule mandating any additional record keeping requirements, Section 204 instructs that the new requirements, when proposed: (1) must relate only to information that is reasonably available and appropriate; (2) must be science-based; (3) must not prescribe specific technologies to maintain records; (4) must achieve public health benefits that outweigh the cost of complying with the requirements; (5) must be practical for facilities of varying sizes and capabilities; (6)

must, to the extent practical, not require a facility to change business systems to comply; (7) must allow for the maintenance of records at a reasonably accessible location, provided that the records can be made available to FDA within 24 h of a request; and (8) must not require a full pedigree, or a record of the complete previous distribution history of the food from the point of origin [4].

When the new rules are in fact proposed, there will be additional clarity on the specific ingredients and foods which FDA considers high risk, and the additional record keeping requirements that FDA will impose on firms which handle, process, and distribute those foods. Although it is believed that once the new provisions are proposed and finalized, the types of available information relating those higher risk foods will be enhanced to some degree, it is also believed that it may ultimately prove difficult for FDA, given the limitations imposed by section 204, to impose any new or additional regulatory requirements which would, in fact, significantly enhance the traceability of those foods.

There are a wide-range of civil and criminal penalties that can result from a company's failure to comply with USDA and FDA traceability and related regulatory requirements. On the civil side, any food company that fails to comply fully with any existing or proposed regulatory requirements can be subject to administrative enforcement actions that include a threat by USDA or FDA to withdraw the company's ability to process and distribute its food products in interstate commerce. On the criminal side, penalties can be far-reaching, ranging from misdemeanor to felony charges. Even where the conduct justifies only a misdemeanor charge (as opposed to a felony charge), the penalties for each count can include up to a year of incarceration and up to $250,000 in individual fines. For regulatory felonies, the penalties can increase to up to 3 years of incarceration per count, and up to $250,000 in individual fines.

Although the existing and proposed rules are in some ways imperfect, both FSIS and FDA do in fact have traceability requirements and enforcement mechanisms on the books. Given the global nature of the supply chain and the disparate food safety systems in place around the world, however, achieving a truly harmonized traceability system for all food ingredients and products is likely a long way away.

References

1. Egg Product Inspection Act, 21 U.S.C. § 1031 (1970)
2. *E. coli* O157:H7 Contamination of Beef Products 67 F.R. 62325 (2002)
3. Establishment, Maintenance, and Availability of Records: Amendment to Record Availability Requirements, 79 F.R. 18799 (2014)
4. FDA's Draft approach for designating high-risk foods as required by section 204 of FSMA, February 2014 [Internet]. Silver Spring: U.S. Food and Drug Administration; c2014 [cited 2015 January 11]. Available from: http://www.fda.gov/downloads/Food/GuidanceRegulation/FSMA/UCM380212.pdf
5. Federal Food and Drugs Act of 1906, 34 Stat. 768 (1906)
6. Federal Food, Drug and Cosmetic Act, 21 U.S.C. § 301 (1938)

7. Food Safety Modernization Act, P.L. No. 111–353 § 204 (2014)
8. Hazelkorn B. The HACCP Epic. Meat Processing, North America Edition [Internet]. c2004 [cited 2016 January11]. Available from: http://www.meatnews.com/mp/northamerican/dsp_particle_mp.cfm?artNum=551
9. McEntire J, Bhatt T. Pilot report for improving product tracing along the food supply system –final report [Internet]. Chicago: Institute of Food Technologists; c2012 [cited 2016 January 11]. Available from: http://www.fda.gov/downloads/Food/GuidanceRegulation/UCM341810.pdf
10. Pathogen Reduction; Hazard Analysis and Critical Control Point (HACCP) Systems; Final Rule, 9 C.F.R. 38806, 38807 (1996)
11. Pathogen Reduction; Hazard Analysis and Critical Control Point (HACCP) Systems, 61 F.R. 38806 (1996)
12. Poultry Products Inspection Act, 21 U.S.C. § 451 (1957)
13. Record Retention Period, 9 C.F.R. 320.3 (1970)
14. Recordkeeping Requirements, 21 U.S.C. § 642 (1967)
15. Recordkeeping Requirements, 9 C.F.R. 416.16 (1998)
16. The Federal Meat Inspection Act, 21 U.S.C. 601 et seq.
17. Records Required To Be Kept 9 CFR 320.1(b)(4)
18. Records, 9 C.F.R. 417.5 (1970)
19. Records, Registration and Reports, 9 C.F.R. Part 320 (1970)
20. Transportation, 9 C.F.R. Part 325 (1970)
21. U.S. Food and Drug Administration. Guidance for industry: FDA records access authority under sections 414 and 704 of the Federal Food, Drug, & Cosmetic Act. c2014 [updated 2015 April; cited 2016 January 11]. Available from: http://www.fda.gov/Food/GuidanceRegulation/GuidanceDocumentsRegulatoryInformation/%20FoodDefense/ucm292745.htm
22. Establishment, Maintenance, and Availability of Records - Definitions, 21 C.F.R. 1.328 (1938)
23. Information Non-transporters Must Establish and Maintain to Identify the Nontransporter and Transporter Immediate and Previous Sources of Food, 21 C.F.R. 1.337 (2004)
24. Information Non-transporters Must Establish and Maintain to Identify the Nontransporter and Transporter Immediate Subsequent Recipients of Food, 21 C.F.R. 1.345 (2004)
25. Information Transporters Must Establish and Maintain, 21 C.F.R. 1.352 (2004)

Chapter 3
Public Health

John Guzewich and Benjamin Miller

Abstract Traceability is intrinsically linked to public health protection. Traceback investigations represent an attempt to reconstruct the food supply chain for one or more foods suspected in an outbreak, with the objective of finding a common location, ingredient, or product. Since it is impractical to assume that a foodborne outbreak will never occur, it is critical to have strong traceability systems in place to rapidly identify the specific food implicated in illness, so that any remaining contaminated product can be expeditiously removed from the distribution system. Regulators must be "fast and right". The pressure to be fast can push agencies to act on less definitive information and therefore with less certainty in the name of protecting public health. The need to be right can result in reluctance to act until more certainty can be had but also possibly resulting in more exposures to contaminated food and more illnesses. When traceability is lacking due to limited recordkeeping, more cases of identified human illnesses are often needed to identify a common grower or manufacturer of the suspect food item in the supply chain in order to collect enough evidence to define a common point of convergence. Improvements in traceability will help regulators be both fast and right, resulting in the greatest public health protection.

Keywords Traceability · Public health · Outbreak · Foodborne illness · Product tracing

J. Guzewich (Retired) (✉)
FDA Center for Food Safety and Applied Nutrition, College Park, MD, USA

B. Miller
Minnesota Department of Agriculture (formerly), Minneapolis, MN, USA
e-mail: ben.miller@achesongroup.com

J. McEntire, A. W. Kennedy (eds.), *Food Traceability*, Food Microbiology and Food Safety, https://doi.org/10.1007/978-3-030-10902-8_3

Introduction

Traceback of foods in response to foodborne illness outbreaks or food contamination events is not a new concept. Investigators linked tuberculosis and Streptococcus outbreaks to raw milk consumption in the late 1800's and early 1900's. Similarly, Typhoid fever outbreaks were linked to the consumption of raw molluscan shellfish, particularly oysters, in the early 1900's. These findings were part of the impetus behind some of the first food safety and traceability regulatory requirements in the US. Shellfish were required to be shipped with a tag on the container that enables tracing back to the harvester and waters where they were harvested. Pasteurized milk containers must show the government issued plant number for the plant where milk was pasteurized. Over ensuing years food safety regulations have required food package labels to include the name and address of the manufacturer, packer or distributor. Food packages and containers often also contain a lot code and or manufacturing date that can further facilitate a product trace.

Table 3.1 identifies several influential outbreaks that shaped food safety regulations, including traceability. In the early 1980's numerous Norovirus and hepatitis A foodborne disease outbreaks were linked to the consumption of raw or undercooked hard shell clams. Inadequate shellfish tagging compliance complicated and slowed the investigation and response to these outbreaks. Later in the 1980's as well as the 1990's shell eggs were linked to many *Salmonella* outbreaks. These outbreak investigations were difficult because the required information on the shell egg containers

Table 3.1 Influential historical outbreaks that involved tracebacks

Influential historical outbreaks that involved tracebacks
Typhoid – raw oysters ~1924
Milk borne outbreak(s) of late 1800s and early 1900s
Raw hard clam outbreaks of 1980's
Shell egg outbreaks of 1980s – 2010
Fresh produce outbreaks of 1990s to present
Listeria monocytogenes in meat outbreaks
E. coli in ground beef outbreaks (Jack-in-the Box)
E. coli in juices (Odwalla and others)
E. coli in cookie dough
Melamine in pet food
BSE in cattle
Peanut butter outbreaks
Botulism in various canned foods
Salmonella in packaged foods like cereal and snacks

indicates where the eggs were packed, but this is not necessarily where the eggs were laid. Since the 1990's there have been many foodborne disease outbreaks linked to the consumption of fresh produce. Fresh produce traceback investigations have been very difficult to accomplish due to the complexity of the fresh produce production and distribution system and inadequate record keeping in the system. Fresh produce usually does not come with source information on it. And when it does that information is generally discarded by the end user. Economically motivated contamination of pet food and infant formula with the chemical melamine in 2007 led to an international investigation. That traceback was complicated by a complex distribution and manufacturing system and inadequate record keeping. These examples provide context for why so much time and effort have been spent on improving the traceability of foods from the point of consumption back to the source or sources of the ingredients and why traceability is so closely linked to public health protection.

When outbreak investigations implicate a food as the means for people to become ill investigators need to determine what that food is and ensure that additional exposure to that food does not occur. They also must determine where and how the food became contaminated. Most reported foodborne disease outbreaks in the US are the result of errors made where the food was prepared. Most often that is at a restaurant. However, if the investigation determines that the contamination did not occur at the place of preparation a traceback is initiated to identify the source. Similarly, if a manufactured food sample is positive for a pathogen as the result of product testing, even if no illnesses are known to have occurred, an investigation and traceback are usually undertaken.

Every foodborne illness outbreak investigation confronts investigators with the "fast and right" dilemma. On the one hand investigators have a legal and ethical sense of urgency to identify the contaminated food as quickly as possible and remove it from sale to prevent any additional illnesses. At the same time they have legal and ethical reluctance to act too quickly and possibly identify the wrong food thereby causing unwarranted economic harm to a firm and its employees. Warning the public to avoid a food that is ultimately determined to not be the contaminated food does not protect them as they still may consume the food that is actually contaminated. Such a mistake also can undermine the credibility of future outbreak investigations in the minds of the public and the industry. The pressure to be fast can push agencies to act on less definitive information and therefore with less certainty in the name of protecting public health. The need to be right can result in reluctance to act until more certainty can be had but also possibly resulting in more exposures to contaminated food and more illnesses.

Collaboration between government and industry is essential for fast, efficient and effective tracebacks, however, some government agencies have been limited in their ability to share certain key information during these investigations by legal restrictions on what they can share. These restrictions have made it difficult to develop trust among governmental agencies and between government and industry. The unfortunate result of these information sharing issues has made it more difficult to protect public health.

Why Tracebacks Are Conducted

Traceback investigations represent an attempt to reconstruct the food supply chain for one or more foods suspected in an outbreak. When multiple people fall ill at multiple locations during the same defined time period, and food is the suspected source, a traceback investigation may be initiated to find "convergence" in the supply chain in an attempt to identify the common source of illness.

Experience with bacterial pathogens and foods likely to be contaminated with these pathogens have given investigators initial leads in some traceback investigations. *E. coli* O157:H7 was commonly associated with ground beef in the 1980s and 1990s until changes in inspection and testing policies pushed to eliminate this pathogen from ground beef. During this time period, a traceback investigation of cases of *E. coli* O157:H7 would regularly begin by tracing the ground beef, or other beef products, consumed by those who became ill in an effort to find a common source in the supply chain. More recently the association between *E. coli* O157:H7 and fresh produce caused investigators to rethink the types of foods associated with this pathogen and start tracing produce exposures more often for cases of illness associated with this pathogen; ground beef was not automatically assumed to be the vehicle. Prior knowledge of a pathogen-food association is not always needed for a traceback investigation. Sometimes a new pathogen-food association is discovered by tracing suspect food items back in the supply chain; as was the case when the first nut-associated *E. coli* O157:H7 outbreak was documented by tracing cases of *E. coli* O157:H7 associated with hazelnuts back to a common distributor and packing house in 2010 [4].

Regardless of a known or unknown association between a pathogen and a suspect food item (or items), one of the primary goals of a traceback investigation is to prevent additional cases of illness from occurring. This requires a traceback to be conducted as rapidly, but as accurately, as possible to find a potential source of convergence in the supply chain. The likelihood of stopping the ongoing distribution of contaminated product is lower when dealing with products with a short-shelf life or high product turnover, such as fresh produce, unless the source of contamination is ongoing (irrigation water, etc.), while the likelihood of stopping the distribution of a product with a long-shelf life, such as peanut butter, is greater.

By identifying the common source of contaminated product in the supply chain, industry and regulators can take effective steps to prevent additional affected product from entering commerce and remove contaminated product from the marketplace through a food recall.

In addition to removing affected product from sale, a traceback also can identify the *source* of the outbreak, allowing regulatory and industry investigators to conduct a root-cause analysis to determine the cause of contamination and prevent similar contamination events, and outbreaks, from occurring in the future. Many of the outbreaks associated with *E. coli* O157:H7 and *Salmonella* in fresh leafy-greens that occurred in the mid-2000s informed the writing of the Produce Safety Rule that stems from the Food Safety Modernization Act (FSMA) of 2011. The Act and resulting rule creates industry standards for water and in some cases product testing, and requirements for manure application, and recordkeeping.

Similarly, much research has been conducted on the effect of time and temperature in the peanut roasting process to eliminate *Salmonella* from peanut butter. These studies have demonstrated that *Salmonella* may be able to survive in higher temperatures and for longer periods of time depending on fat and water activity of the food matrix in which they are present [2].

Within the regulatory and public health community, a distinction has arisen between the "type" of traceback investigation that may be conducted by a regulatory agency. When the epidemiological investigation fails to definitively identify a single suspect food item associated with cases of illness in an outbreak, an "investigational" or "epidemiological" traceback may be conducted on several food items to determine if one of the items demonstrates convergence in the supply chain that can explain the majority of illnesses associated with the outbreak. This type of traceback is conducted with a focus on speed, and records may be collected by email, phone, or fax in an attempt to rapidly reconstruct the supply chain and find a point of convergence. "Epidemiological" tracebacks have been used with good success by a number of state regulatory agencies and played a key role in determining the actual source of illness in the nationwide *Salmonella* Saintpaul outbreak originally associated with tomatoes. The "epidemiological" traceback in this investigation ultimately identified jalapeno and Serrano peppers imported from Mexico as being the likely source of the outbreak.

Unlike an "epidemiological" traceback, a "regulatory" traceback is typically started when there is a strong indicator that a single food item is associated with illnesses in an outbreak. The level of evidence collection in a "regulatory" traceback is usually greater, with federal or state investigators typically visiting each facility that handled the suspect product in-person to conduct a focused investigation of the facility and determine if other plausible sources of contamination could explain the cases of illness in the outbreak (other than, or in addition to, the suspect food item). For example, if the suspect food item was stored in a warehouse prior to distribution to area grocery stores, could there have been a source of contamination within the warehouse that could have affected the food item and be the plausible source of the outbreak?

The difference between epidemiological and regulatory tracebacks represent the inherit tension between the aforementioned "fast and right" dilemma; if contaminated product still exists in the marketplace, what is the obligation of the regulatory and public health community to move with the utmost speed to identify the source of the outbreak? Does the benefit of testing all possible hypotheses in a regulatory traceback in an effort to build an airtight enforcement case outweigh the immediacy of preventing ongoing public exposure to contaminated product? These are the questions that are debated during many traceback investigations between public health and regulatory agencies in an effort to find the balance between being "fast" and being "right". Similarly food industry representatives argue that regulatory action or public health alerts about the possible source of contaminated food should be delayed until a traceback can definitively link an illness outbreak or food contamination to as specific a source as possible in order to limit the scope of adverse impact on an entire food category. Once again government agencies are balancing "fast" and right."

Description of a Traceback

A traceback, in the purest sense, is simply an extension of the epidemiologic investigation initiated by local or state epidemiologists and represents an effort to further and more accurately characterize an exposure to a suspect food item for each case of illness in an outbreak. A traceback investigation typically begins after epidemiologists have interviewed cases of illness in an outbreak and narrowed down the number of suspect food items.

At this point, state or federal food regulatory investigators work with the epidemiologists to identify the most likely location(s) of exposure for each case in the outbreak and then collect product distribution records from these locations. For example, someone may have purchased a contaminated food item from a grocery store while another person may have consumed the same contaminated item from a restaurant in a different city or state. In an investigation, each of these locations where a person was "exposed" to the suspect food item represents a "leg" in the traceback investigation which represents a unique location where one or more cases were exposed to a suspect food item. A leg with multiple cases of illness occurring at the same or similar times is considered a better leg to trace back since it is more likely that contaminated food was sold at this location due to multiple cases of illness. A single illness within a leg can still yield valuable information for an investigator but may be considered of lower evidentiary or epidemiological value.

When initially investigating a leg in a traceback, regulators will collect invoices and purchase order information from each of these locations and start to trace the suspect food item(s) back through the supply chain. In most instances, food delivered and sold to retail locations such as grocery stores or restaurants typically traces back to a food distribution company with a warehouse that may be located nearby or several states away.

While atypical, an investigation may identify a local source of suspect product that has been delivered directly from a food manufacturer or local farm. In these instances, the traceback leg is considered "short" since there are only one or two steps between the source of the food and the end consumer.

The Bioterrorism Act of 2002 and FSMA created recordkeeping requirements for most companies in the food supply chain (mostly exempting farms and retail locations). Based on these recordkeeping requirements, most food producers and distributors maintain adequate records of food transactions between external entities in the supply chain (their suppliers and their customers) as illustrated in Fig. 3.1. These records have existed and been available to regulators conducting tracebacks largely because they document transactions and facilitate payment

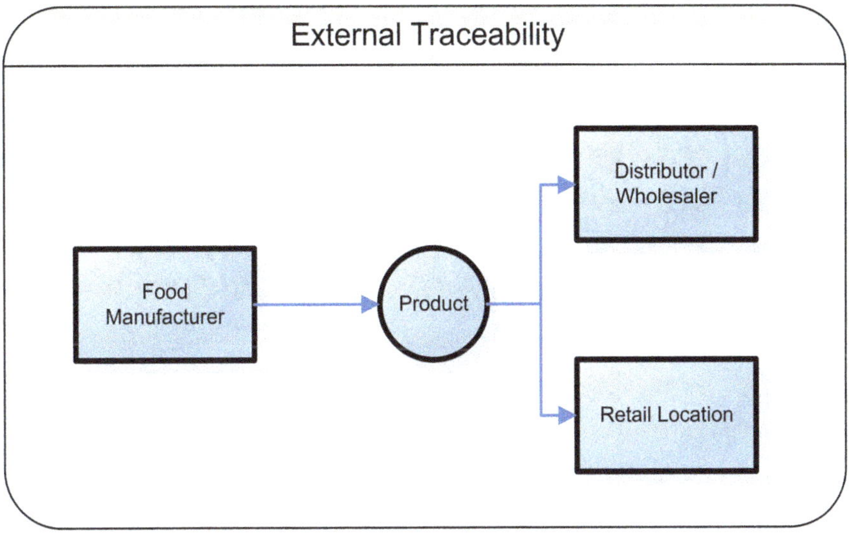

Fig. 3.1 Schematic of external traceability

between companies in the food supply chain. Investigators usually refer to these documents, which include invoices, purchase orders, and bills of lading, as maintaining "external" traceability within the supply.

Investigators can use these "external" documents to trace the suspect food item from the distributor's warehouse back to a manufacturing facility where the food item was created using a large number of ingredients.

Traceback investigations often encounter challenges when trying to establish "internal" traceability within a food distributor or manufacturer, shown in Fig. 3.2. Until recently, many food distribution companies and food manufacturers did not have processes or data systems that would maintain "internal" traceability in the supply. Internal traceability requires that food processors or distributors track internal inputs that change the identity or configuration of the product they are selling. For food manufacturing, internal traceability may require that all lot code or batch information for the ingredients (grain, corn syrup, flavorings, vitamins, etc.) that are used be recorded and stored. For a distributor, internal traceability may require that multiple data elements be recorded if cases of product from varying lots are used to create a pallet (or an equivalent logistical unit).

If internal traceability is not accurately or consistently maintained within a facility or for the food item of interest in an investigation, complete supply chain traceability is lost. This loss of internal traceability frequently occurs when dealing with manufactured foods with multiple ingredients, bulk foods, and fresh produce items that are not packaged.

In order to continue a traceback investigation in these instances, investigators must infer the likelihood that a suspect product was shipped from or received at a facility based on temporal bracketing that takes into consideration the throughput of

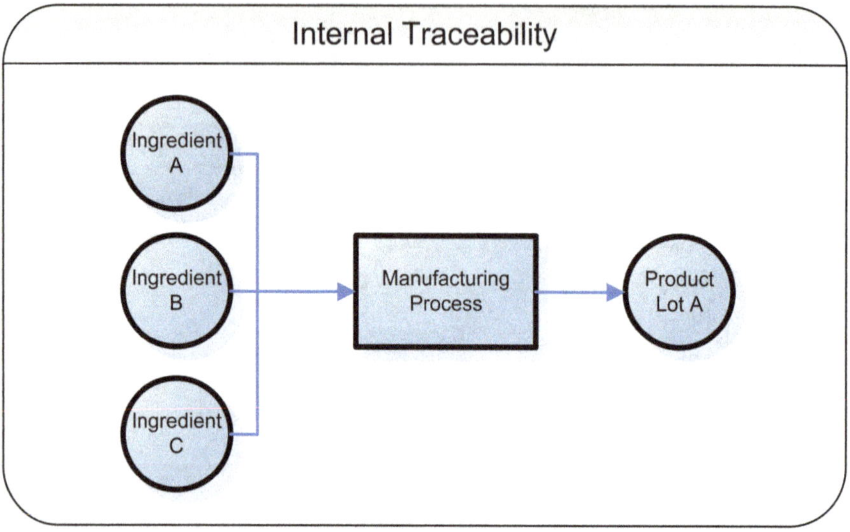

Fig. 3.2 Example of internal traceability

the product in that facility. For example, if a fresh produce item typically is stored for 2 days in a cold storage warehouse, the investigator may ask for product receipt and shipment documentation for a week on either side of the date of interest for the suspect food item being traced. This temporal bracketing should capture all suspect transactions (both receiving at the warehouse and shipments from the warehouse) in an attempt to identify possible suspect sources of the contaminated product.

In investigations where internal traceability is lacking due to the specific food and supply chain relationship, several traceback legs that cover multiple distribution channels are often needed to identify a common source of convergence in the supply chain. Simply put, due to limited recordkeeping, more cases of identified human illnesses are often needed to identify a common grower or manufacturer of the suspect food item in the supply chain in order to collect enough evidence to define a common point of convergence.

An Example Investigation

Traceability should cover the entire food supply chain; from animal feed to finished food products regardless of risk classification. Numerous foodborne outbreaks over the past several years have demonstrated the importance and need for rapid traceability of food products sold to consumers. More rapid traceability can aid and clarify foodborne illness investigations by aligning product distribution data with epidemiological exposure data. These investigations could be completed more rapidly and with a greater degree of accuracy if current data requirements and

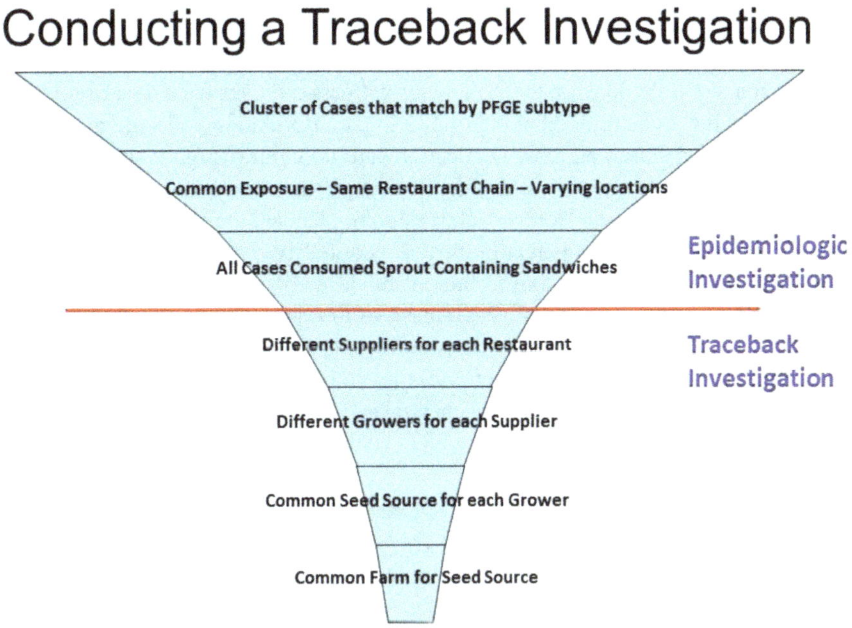

Fig. 3.3 Stages of a traceback investigation

collection practices for food traceability were better defined and aligned across the food supply chain.

Figure 3.3 shows a cluster of illnesses matching a Pulse Field Gel Electrophoresis (PFGE), genetic fingerprint, subtype identified by epidemiologists at a state health department. Due to the difficulty and cost of traceback investigations, significant clusters of illness must be identified prior to an investigation commencing. Once the cluster has been identified, epidemiologists interview specific cases and try to determine commonality such as dining at operations of the same restaurant chain. Further investigation finds that all cases consumed some type of sprout-containing sandwich at each restaurant location but the source of the sprouts remains unclear.

While the epidemiological investigation may identify a plausible source, the regulatory investigator must trace the likely exposure to a point of convergence or commonality in the supply chain in order to identify the confirmed "source" of the outbreak.

Continuing with the example, once an outbreak vehicle is identified, the epidemiological investigation ends with the possible recommendation that persons not consume sprout-containing sandwiches at these locations. The traceback investigation is an extension of the epidemiological investigation and it serves two purposes:

1. It supports the epidemiological associations by confirming that temporal and physical distribution of suspect products could adequately match the case exposures, and;

2. It further characterizes the source of the outbreak, thereby increasing the likelihood of a meaningful intervention to protect public health.

This concept is demonstrated when the investigation moves from the information generated in the epidemiological interviews to the information collected by a food-regulatory agency based on record collection and in-field investigations. An investigation of the invoices and bills-of-lading from each restaurant location where a case of illness was reported shows that each restaurant received their sprouts from a different supplier. Further investigation of the records from the suppliers shows that they received their sprouts from a number of different growers. A review of the grow-logs, seed sources, and invoices at each of the sprout grower locations shows that all of the seed in implicated time frame would have come from a single seed supply company. A review of the lot-codes for the implicated seeds shows a common lot-code of seed was used at each grower in the implicated time frame and further investigation shows that the lot-code corresponds to a single farm that produced all of the questionable seed.

It isn't until all of these data are collected and analyzed that a truly meaningful public health intervention, in the form of seed and sprout recalls and a market withdrawal of the implicated lot-code, can be made. As might be imagined, such an investigation is complicated and time and resource intensive. Often, outbreaks subside before investigators are able to pinpoint a cause resulting in wasted time and effort of on the parts of both those doing the investigating and those being investigated.

Epidemiology

Epidemiology is the study of diseases in populations. Public health agencies use epidemiology to understand what the public health problems are so that appropriate prevention activities can be undertaken. Epidemiologists use various surveillance approaches to determine what diseases are occurring. For many foodborne diseases, health providers and clinical laboratories are required to report cases of foodborne illness to health departments. The agencies follow-up to prevent further spread of those diseases as well as determine how they were spread. If contaminated food is determined as the cause of the infection epidemiologists and food regulatory staff work together and alongside industry to prevent additional exposures through public alerts and recalls. They also investigate to determine what factors contributed to the contamination of the food. If a food source is determined to have been where the food was contaminated a traceback investigation will be conducted to identify that source.

Most foodborne disease outbreaks in the US are identified by consumer complaints. In these cases epidemiologists, laboratories and food regulatory officials investigate as described above. Detection and investigation for foodborne illness outbreaks have improved in recent years thanks to improvements in laboratory methods to detect and link bacteria and viruses recovered from people, food and the environment. Increased government support for epidemiology and laboratory testing and improvements in collaboration and cooperation between government agencies at the local, state, and federal levels have also led to improved detection, investigation, and response to foodborne disease outbreaks.

Typical steps in foodborne outbreak epidemiology are surveillance, detection, investigation and response. The surveillance step involves epidemiologists and laboratories looking for "clusters" of "sporadic" cases of diseases like salmonellosis. Clusters refer to two or more ill persons, or cases, with commonalities in who is involved, where they have been or the time frame of onset of their illness known as person, place and time associations. Consumer complaint investigations constitute the other common surveillance method for identifying food outbreaks. Once a cluster of cases or a consumer complaint has identified a possible outbreak this constitutes detection. Investigators then interview ill and similarly- exposed well people to identify possible specific exposures that were more likely to have occurred in ill people when compared to the well people. If the more common exposure for the ill people has identified a common food the investigators then respond by acting to prevent further exposures to that food. They also investigate how the food became contaminated so that future outbreaks can be prevented. Investigating agencies are often working with incomplete information during their investigations. In some such cases epidemiologists may advise regulatory officials that in their judgment their epidemiological information is sufficiently compelling that regulatory agencies need to act before additional information is gained in order to minimize the public health impacts of the outbreak. These all too common situations were described above as the "fast and right" dilemma.

As surveillance and detection have identified more outbreaks they have been discovering outbreaks that are more difficult to solve. These difficulties can best be illustrated by so called ingredient-driven or stealth ingredient outbreaks. In these investigations the suspected food is not something obvious like raw milk or an inadequately cooked turkey. Instead the suspected food is an ingredient served at a restaurant with all menu items having multiple ingredients. In such cases epidemiology cannot differentiate between all of the ingredients in, for example a salad or a taco. Investigators therefore face the need to trace not one ingredient but many ingredients in hopes that the tracebacks will identify the contaminated ingredient. As more and more traceability systems are implemented these sometimes-unsolvable outbreaks may become solvable as investigators should be able to quickly identify the source(s) of all ingredients in a complex food and from that help to point to the most likely contaminated ingredient. This approach will work best in those instances where outbreaks are occurring at multiple sites, and tracebacks from the multiple sites only converge at one point, for one ingredient, while traces on the other ingredients do not converge. This approach has worked in the past as well, but because of the many challenges that exist to traceability at the current time the traces are very resource intensive and slow if they work at all.

Laboratory

Several types of laboratories play key roles in supporting epidemiological and traceback investigations. Most states maintain a centralized public health laboratory where bacterial isolates obtained from ill persons (typically isolated from clinical samples of stool, urine or blood depending on the pathogen) are received and analyzed using advanced molecular methods. Since the mid-1990s, Pulse Field Gel Electrophoresis (PFGE) has been a widely used method to identify the DNA "fingerprint" of bacteria associated with an outbreak. These "fingerprints" are uploaded by the public health laboratories to a central database (PulseNet) maintained by the Centers for Disease Control. As described below, additional fingerprinting methods such as Whole Genome Sequencing have recently been employed during foodborne illness investigations. Epidemiologists use this information to link human cases that may be associated with a common outbreak.

Other laboratories, typically associated with food regulatory agencies at the state (departments of health or agriculture) or federal level, specialize in isolating pathogenic bacteria from suspect food items using similar molecular methods. Unlike clinical isolates, these labs must coax small numbers of the bacteria from a wide variety of food matrices and isolate viable bacterial colonies using methods that have been in place for decades. Once a suspect bacterial colony has been identified, a variety of molecular methods can be used to determine if the bacteria found in the food item matches the bacteria common to human cases in the outbreak. PFGE can be used to determine if there is a "match" between the food and human cases of illness. When a link between a suspect food item and human illness is confirmed by a matching PFGE subtype (or increasingly, whole genome sequencing), this represents the "gold-standard" in an outbreak investigation and the traceback may begin or continue in earnest to identify the ultimate source of the food item that has tested "positive" for the outbreak strain. At this point, if the source of the product remains unclear (bulk produce vs. a commercially manufactured food product with a lot code on the packaging) the traceback investigation will move quickly to identify the source (farm, packing shed, etc.) of the food item and determine if other contaminated product may still be in the marketplace or in consumers' homes.

Laboratory results and the expectations that both the food industry and regulatory agencies place on obtaining a positive food product sample represent another illustration of the "fast and right" dilemma; often both industry and regulators would prefer to substitute laboratory evidence for traceback evidence in an investigation since a positive lab finding removes all doubt of the association between consumption of the food product and subsequent illness. The reality remains that obtaining a representative food sample and isolating the outbreak strain of bacteria from this sample during an investigation can be extremely challenging since the product may no longer exist or it may be a in a condition that makes the likelihood of isolating viable bacteria very low.

Ideally, laboratory results from food samples should be used to support the trace-back hypotheses developed by investigators during an investigation and record collection. Analysis should occur concurrently with sampling and laboratory analysis.

As laboratory technology advances new molecular methods are being used to identify clusters of human illnesses and link these to positive food or environmental samples. Current PFGE methods require a bacterial culture in order to conduct the analysis and generate a "fingerprint". Clinical laboratories, for reasons of cost and rapid diagnostics, are shifting away from culture-based methods of pathogen identification and moving to non-culture-based methods.

This shift to non-culture-based clinical methods is currently challenging the Center for Disease Control (CDC) and PulseNet to identify new culture independent methods for genetic fingerprinting and cluster identification. New methods may include metagenomic analyses that sequence select genetic targets or all bacteria in a patient sample and could identify bacteria making a patient sick and subtype information at the same time. Whole genome sequencing is increasingly being relied upon as a more discriminatory technique to definitively match patient and food and/or environmental samples. Table 3.2 shows a comparison between Next Generation Sequencing technologies and the current PFGE "Gold-Standard" approach [1].

Regulatory

Regulatory responsibility for the safety of the food supply is shared between three levels of government in the US. The federal government, primarily the Food and Drug Administration and the Food Safety and Inspection Service of the Department of Agriculture, have jurisdiction over foods or ingredients that have traveled in

Table 3.2 Comparison between molecular techniques to investigate outbreaks [1]

Key factors	Next Generation Sequencing (NGS)	Current "Gold Standard" (PFGE)
Time to get data	1–2 days	Days
Time to analyze data	Hours to days	Hour or less
Cost per bacteria isolate (for reagents only)	~$150–300 (depends on platform)	~$10–15
Ability to differentiate isolates	To be determined but likely very good	Well-established for bacteria tracked by PulseNet, few exceptions
Potential for automation	Yes	No
Intended objective	Genetic serotyping, virulence, antimicrobial resistance profiling, and subtyping possibly based on SNPs or other variant regions	Only subtyping

interstate commerce. State agencies such as Departments of Health and/ or Agriculture and/ or Environment and/ or Inspections have jurisdiction over food safety within their respective states. Local, county or city, government usually has jurisdiction over restaurants and food markets within their jurisdiction. These jurisdictions vary by state and locality and there often are overlapping authorities at two or more levels of government. These complexities can make food safety regulation a confusing thing to understand and to carry out. These layers of government operate under a multitude of laws, rules, and regulations and each agency can have its own set of priorities. Many of the agencies have inadequate resources to carry out their respective responsibilities. When regulatory agencies are called upon to assist in an outbreak investigation they must assess their response in light of other conflicting priorities and limited resources. If they judge the effort to be too great or complex in light of their other realities a slow and/ or inadequate response may result. A slow or inadequate response for one agency in one part of the country could therefore result in inadequate public health protection in another part of the country. The roles that regulatory agencies play in outbreak surveillance and response can also vary widely. In some places regulatory staff can be involved in surveillance and detection activities like following up on sporadic cases of disease or interviewing ill and well persons in the investigation. The most common role of regulatory agencies is in carrying out investigations at the place(s) where food was prepared and/ or contaminated and in conducting traceback investigations. Regulatory agencies can also direct that product recalls be conducted, they can issue press releases alerting the public to food safety risks and they can initiate enforcement actions when appropriate.

Challenges in Tracebacks

If used as a tool to test an epidemiological exposure hypothesis, traceback investigations should be initiated for most outbreaks assumed to be associated with a commercially distributed food. While not the current practice in all jurisdictions in the United States, this liberal application of the traceback investigation as an outbreak tool will ultimately identify the source of more outbreaks than if tracebacks are conducted conservatively. However, in either case, for every traceback investigation that is started and succeeds, there are at least equally as many investigations that never find convergence in the supply chain or identify the source of an outbreak. There are myriad reasons that a traceback can stymie an investigator, both from a regulatory perspective (Table 3.3a) and epidemiological perspective (Table 3.3b).

Poor recordkeeping is often the cause of an unsuccessful traceback investigation. Despite the recordkeeping requirements of the Bioterrorism Act of 2002 and the forthcoming FSMA requirements, it is evident that many food businesses fail to maintain required records for one-step traceability (the ability to document where

Table 3.3a Traceback challenges for regulatory agencies

Traceback challenges for regulatory agencies
Competing priorities
Varying resources and expertise
Differing regulations and authorities
Global food supply
Need to be fast and right
Large numbers of background sporadic cases
Poor consumer recollection of consumption history and lack of specific product information
Multiple product varieties identified
Multiple products with multiple ingredients identified
Produce: perishable
Lack of rapid connectivity
Lack of unique identifier
Lack of interoperable systems to rapidly link product from farm to fork
Legal restrictions on information sharing

Table 3.3b Traceback challenges for epidemiologists

Traceback challenges for epidemiologists
Ingredient-driver/stealth ingredient outbreaks
New vehicles not previously reported
Differing regulations and authorities
Differing interview forms used in different agencies
Clinical labs moving to culture-independent diagnostic tests
The need to be fast AND right
Loss of laboratory and epidemiology resources in some state and local agencies
Coordination of large multi-agency outbreak investigations
Lack of uniform and in-depth follow-up of all sporadic cases limits detection
Need for improvements in electronic information sharing between public health labs and epidemiologists

product is received from and where outgoing product has been shipped). Business records such as purchase orders, bills of lading, invoices, and production records often do not contain all the relevant information that an investigator may be seeking. These documents are often copied, scanned and faxed many times throughout the supply chain and contain important, and often illegible, handwritten notes or comments about delivery dates and times. In additional to missing or illegible records, products (especially bulk produce items) are often referred to by different names by each company handling them in the supply chain. For example, a tomato grown in Florida may start its existence in the supply chain as a "pink #4 tomato", travel to a distributor who calls the same tomato a "#4 red tomato" and end it journey as a "red slicer" on the restaurant chain invoice. This variable nomenclature for the same product can cause confusion and delays for investigators trying to determine if they're still tracing the same product.

Products with a short shelf life such as fresh produce and seafood can be challenging to traceback through the supply chain as well given that these products are not often packaged and are sold in bulk or cases with minimal lot code information associated with the food product. Furthermore, if lot code information is included, most food distributors still do not have the capacity to track lot code specific information at the "case" level within their warehouse. Throughput of product is largely handled on a "first-in-first-out" basis. This lack of case-level traceability truly hamstrings investigators when trying to determine to what retail locations product was shipped from the distribution warehouse. Additionally, short shelf life products are often out of the marketplace and may have been fully consumed by the public prior to a traceback investigation even beginning. Gaining the support of regulatory agencies to conduct traceback investigations when the risk of additional illness appears minimal can be difficult and the incentive of conducting a root-cause analysis or environmental assessment to better understand the cause of the outbreak, at times, isn't always compelling enough to conduct an investigation.

Packaged food items with a long shelf life can present another challenge to investigators since companies may only maintain records for 3–4 months after a transaction has been completed. For a product with a 12-month shelf life, this means that information about the timing of shipments for these products may be unknown to investigators due to a lack of available records. In most instances packaged foods are easier to trace than non-packaged foods but if an outbreak is caused by a contaminated ingredient in a packaged food, such as peanut paste used in a more complex product, many packaged foods may be affected and therefore a common point of convergence in the supply chain can be difficult to identify unless an investigator can get back to the point where the common ingredient was manufactured.

Most food companies are familiar with the concept of conducting a food recall and most, erroneously, assume that conducting a traceback investigation is an analogous activity. When recalling product, a firm is simply identifying all of the locations where they shipped the affected product and these records are readily available within their own data systems. A firm can be cautious and recall more product than truly may be contaminated. Contrast this with a traceback investigation where investigators are trying to identify a limited number (most oftentimes *one*) suspect shipment into a facility. The desired specificity in a traceback investigation far exceeds the specificity needed in a recall and most companies struggle to review all the relevant records and identify a limited number of suspect shipments (see internal traceability).

The meat and poultry industry is highly centralized in the United States, with a handful of companies maintaining highly integrated vertical supply chains which can make traceback investigations of meat and poultry difficult because most of the population is exposed to products delivered through one of these companies. When a supply chain for a particular product lacks "diversity" a traceback investigation may identify a source (processing plant for example) that is common to all cases of illness, but since pathogens like *Salmonella* are not considered as adulterants in meat or poultry, a root-cause investigation seems unnecessary.

Example of a Successful Traceback

A successful traceback is one that was both fast and right and as a result the public's health was protected and economic harm to other companies was minimized. On Friday, June 28, 2013 FDA was notified by CDC of a cluster of four *Listeria monocytogenes* cases with the same Pulsed Field Gel Electrophoresis (PFGE) genetic fingerprint. As described earlier, PFGE is a laboratory test that can suggest the relatedness of bacteria from two or more different samples. One of the cases involved an adult death and a second case involved a fetal loss. Based on their preliminary information CDC also reported that soft cheese might be the possible contaminated food. FDA checked their records for past reports of this bacteria having been found in tests they had performed on foods or food processing environments. Several of these results were from a certain soft cheese manufacturer. This firm had a history of having this specific organism in its processing environment. Further, the products manufactured by the firm had been distributed in the areas where the four reported cases resided. By Monday, July 1 follow-up by the states had identified that at least three out of four of the cases had direct or indirect exposure to cheese from the soft cheese manufacturer. An intensive investigation at the plant began on July 2 and on Wednesday, July 3 one state reported that they had preliminary positive *Listeria* findings from an unopened sample of the suspect cheese. The company issued a voluntary recall of the cheese on July 3, 2013.

Case Study of an Unsatisfactory Traceback Investigation

On Friday May 22, 2008, the Friday before the 3-day Memorial Day holiday weekend, the New Mexico Department of Health notified CDC, and subsequently CDC notified FDA, of a growing cluster of *Salmonella* Saintpaul cases occurring in New Mexico. Over the holiday weekend the New Mexico public health laboratory worked to test additional specimens they had received from clinical laboratories in the state and New Mexico epidemiologists began an investigation to determine a common exposure for the growing number of cases being reported. *Salmonella* Saintpaul was a relatively unusual type of *Salmonella* in the US prior to this outbreak, having been reported in about 400 persons per year in the country. The specific PFGE pattern for this strain of *Salmonella* Saintpaul had been seen about 25 times per year in the country. In the previous year six cases had been reported nationwide between April 1 and June 30. The New Mexico public health lab had already identified seven cases of *Salmonella* Saintpaul in May 2008, four of which were of the very unusual PFGE pattern. On May 23 CDC learned of three additional cases of *Salmonella* Saintpaul with the same PFGE pattern being reported by Colorado and Texas. There had been 19 confirmed or suspected reported outbreaks linked to *Salmonella* Saintpaul in the time period 1998–2006 in the US. Foods that were reported as possible vehicles (the contaminated food that was eaten) in these outbreaks included turkey, sprouts, mango, beef, eggs, infant formula, unpasteurized orange juice, tomato, chicken and lettuce. Epidemiologists who were interviewing ill and well persons included the foods historically associated with *Salmonella* Saintpaul in the list of possible food and other exposures that the interviewees were asked about.

By early the next week the number of reported cases of illness continued to grow and CDC notified FDA that the epidemiological investigation was implicating tomatoes as the likely contaminated food. The number of reported cases of illness grew every day with the same states reporting more cases and new states reporting cases (Fig. 3.4). The epidemiological link to tomatoes continued to be reported as well. At this point the epidemiologists only had information on reported tomato consumption. They did not know the type of tomatoes the people had eaten which made it difficult to narrow down the scope of the problem. CDC hosted daily conference calls that included epidemiologists from a growing number of states as well as FDA. Based on the rapidly growing number of cases of *Salmonella* Saintpaul infection being reported, particularly in New Mexico and Texas at this point, on June 3, 2008 FDA issued a Consumer Advisory to people in those two states not to eat Roma or plum tomatoes.

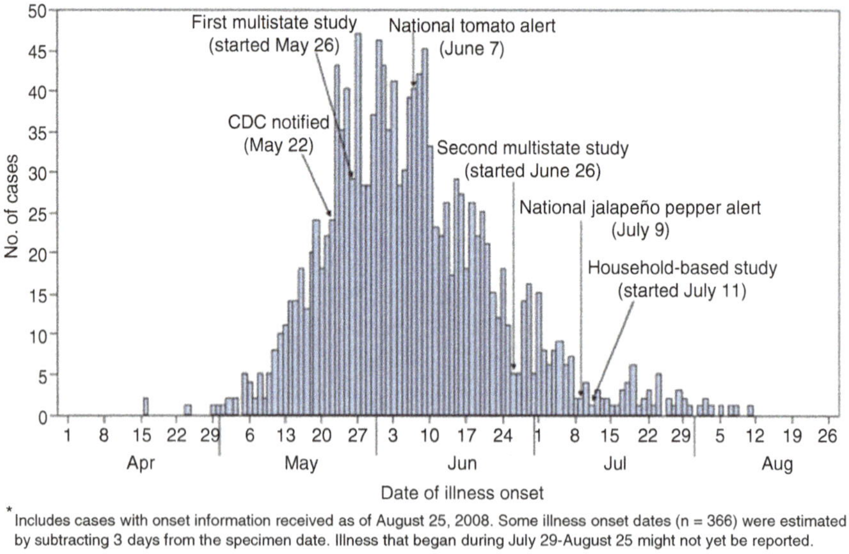

Fig. 3.4 Number of laboratory-confirmed cases (n = 1414) of Salmonella Saintpaul (outbreak strain), by date of illness onset—United States, 2008* [5]

FDA requested that CDC and the states provide information on a number of the ill people who reported having eaten tomatoes so that traceback investigations could be initiated to try and determine the source of contaminated tomatoes. FDA would have preferred to trace from a common point of exposure such as one restaurant or one supermarket, but up to that point in the investigation there were no known clusters of illness that were linked to such a location. Investigators have greater confidence when two or more people have a common exposure, like tomatoes eaten at a common restaurant, compared to a one person's reported exposure. Lacking that information FDA needed single cases who had only one known tomato exposure and who were very certain where their exposure had been. If a case had multiple tomato exposures or did not know where their exposures were, tracing would have been very difficult if not impossible. CDC provided FDA with names of the best cases they could identify for tracebacks. These cases were in different parts of the county with the idea that if the source of tomatoes for multiple cases in multiple states converged at one point, (e.g., one farm/packer) a strong case could be made that was the source of contaminated food. FDA and state regulatory staffs then visited the reported points of exposure and began tracing tomatoes back to their sources. The goal of these traces was to quickly identify the source of contaminated tomatoes so that the public could be given more specific information about what tomatoes to avoid, to initiate a recall and to determine how the tomatoes were becoming contaminated. Such an outcome would also limit the economic hardship that the broad consumer advisory was creating to tomato producers that were not identified by the traceback.

FDA immediately began reaching out to the tomato industry to determine likely sources of tomatoes in the US at that time of the year as well as the extent of distribution of those tomatoes. Florida and Mexico were the largest supplying regions of tomatoes. One industry report was that Florida tomatoes were not likely to be found west of the Mississippi due to shipping costs. That information was soon disproven when a case in Idaho was found to have eaten tomatoes from Florida. By June 7, 2008 the number of cases of illness was continuing to grow as was the number of states across the country that were reporting cases. As a result, on that day FDA issued a nationwide consumer advisory to avoid eating red round or Roma/plum tomatoes.

The traceback investigations were proving to be challenging for a number of reasons including that some tomato packers were packing tomatoes from multiple growers in one packing house, repacking and sorting of tomatoes from multiple suppliers one or more times in the distribution system, renaming the tomatoes from one distributor to the next (6's, to red round, to beefsteak, to cookers), and generally poor record keeping (missing information, illegible information, information that did not add up from one point to the next) that made it next to impossible to determine where tomatoes came from. In spite of these problems it was becoming apparent that tomatoes were coming primarily from a few large packers in Florida and Mexico. This by itself was not surprising as these large packers were the major suppliers at that time. Therefore, this information in and of itself could not be interpreted as implicating these packers as the source of contamination. FDA arranged for teams of investigators to visit these packers as soon as possible.

Investigators also were learning that the lack of convergence in the tomato traces for the cases being traced were shown to in part to be due to the fact that they had more than one tomato exposure, contrary to what had initially been reported. This meant that FDA could have been tracing the incorrect tomatoes even if tomatoes were the contaminated food. On June 19 CDC and the states began intensive investigation of a few clusters of multiple cases of illness that were linked to common exposures such as restaurants in hopes that traces from such locations could be done with more confidence that the right food was being traced.

Many investigators at this point were becoming concerned that tomatoes may not be the contaminated food in the outbreak. Many of the single cases as well as some of the clusters were associated with eating Mexican foods at restaurants. Many of these menu items have multiple ingredients that had the potential to be contaminated with *Salmonella.* FDA and many state laboratories had been testing fresh tomatoes with no samples finding the outbreak strain of *Salmonella*; now these laboratories began testing cilantro, peppers and basil as well. The FDA teams that had visited farms and packers in Mexico and Florida completed their visits after having no significant observations. The many samples of tomatoes, water, soil and other environmental specimens collected during these investigations did find a few *Salmonella* positives, but none matching the outbreak strain.

In late June the Minnesota Department of Health reported having investigated two clusters of *Salmonella* Saintpaul that were strongly associated with having eaten Jalapeño peppers. The Minnesota Department of Agriculture had conducted a preliminary traceback for these peppers which led to an importer in McAllen, TX and from there into north eastern Mexico. FDA immediately sent an inspector to that importer to sample incoming Jalapeno peppers. FDA also sent an investigation team to Mexico to continue the pepper traceback and to visit packinghouses and growing fields. The FDA investigation team in Mexico had to cut their visit short due to an approaching hurricane but before they left they were able to narrow the traceback to a few possible growing fields which they visited and where they collected samples of water, soil and peppers. On July 9 FDA issued a Consumer Advisory to avoid consumption of Jalapeno peppers. On July 21 FDA announced that the outbreak strain of *Salmonella* Saintpaul had been recovered from the sample collected from the importer in McAllen, TX. Subsequently, FDA identified the outbreak strain of *Salmonella* from a Serrano pepper sample and from water samples obtained from the Mexican farm most strongly implicated by the traceback (Fig. 3.5).

In summary, early epidemiological studies implicated tomatoes as the contaminated food in this outbreak investigation. Subsequent tracebacks, farm, and packinghouse investigations as well as microbiological testing of tomatoes did not support the epidemiological findings. The tomato industry lost many millions of dollars, some businesses went out of business and many employees lost income, if not their livelihoods. Further, the public continued to be exposed to the risk of eating Jalapeno and Serrano peppers as they were advised not to eat tomatoes. In terms of

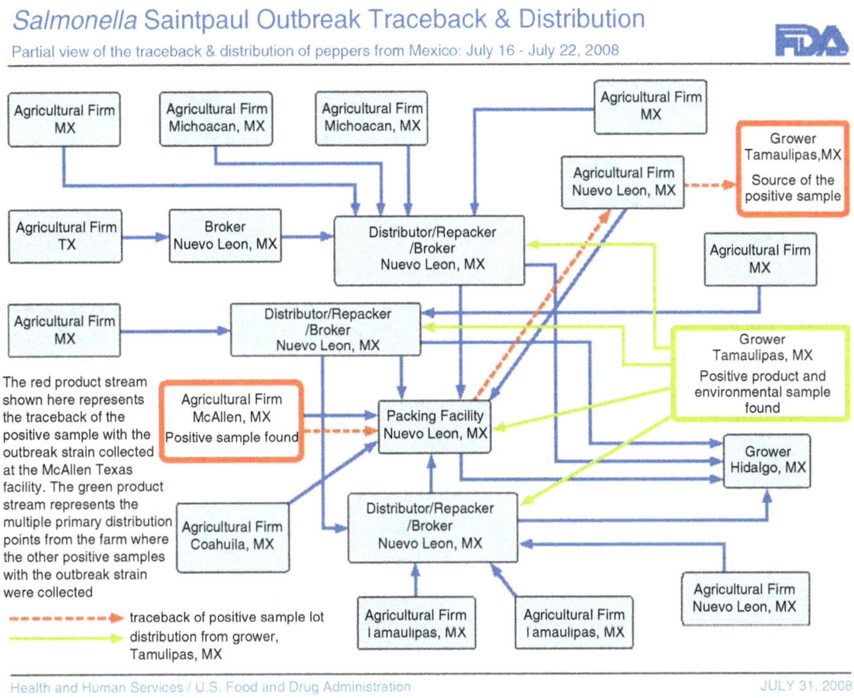

Fig. 3.5 Salmonella Saintpaul outbreak traceback & distribution partial view of the traceback & distribution of peppers from Mexico [6]

traceback lessons learned, government and industry saw firsthand the extent of a crisis that could have been much more limited if fast and definitive tracebacks on tomatoes could have demonstrated in early June that there was no common source of tomatoes to account for this outbreak. In addition, such a traceback system could have shown that other foods served in Mexican restaurants, namely Jalapeno and Serrano peppers being served in multiple states did have a common source.

Public Health Benefits of Traceback Investigations

The Food Safety Modernization Act charged the FDA to conduct a pilot study of ways to improve traceability in the food supply chain using integrated data systems and specified additional recordkeeping requirements for food businesses. Faster traceback investigation depends, in part, on the food industry collecting and maintaining more information in a format that is easily accessible and can be rapidly analyzed by investigators. An outcome of this pilot study proposed a process-based approach to traceability, similar to the Hazard Analysis Critical Control Points process already used by the food industry, called Critical Tracking Events (CTEs) and

Key Data Elements (KDEs) [3]. This flexible traceability model allows food businesses to identify key points in the harvesting, manufacturing, and distribution process where "key data" are consistently captured and stored for the shelf-life of the product. This more systematic approach to data collection will make conducting traceback investigations faster and more accurate, thereby helping to alleviate a portion of the "fast and right" dilemma.

Improvements in data accuracy and collection will aid in shortening traceback investigations, thereby limiting the number of illnesses that may otherwise have occurred once an outbreak has been identified. In addition to improvements by industry, the public health surveillance system at the national, state and local levels need to be strengthened as well. Foodborne illness surveillance is conducted very differently from state to state with greatly varying levels of resources and expertise associated with outbreak investigation. Almost every traceback investigation (except those initiated due to a positive food sample), relies on sound epidemiological data to guide the initial stages of the traceback investigation.

Likewise, regulatory agencies pursue leads and conduct traceback investigations differently, depending on the number of dedicated resources they can commit and on their level of expertise at conducting these investigations. Recently, FDA has created a course for regulators on how to conduct effective traceback investigations and has been training FDA and state regulators in these methods.

When the industry, public health and regulatory agencies are "fast and right" in identifying the source of an outbreak, the public's confidence in the food supply is maintained and industry segments are protected against commodity-wide consumer advisories. If a consumer advisory is necessary, a rapid and accurate traceback can identify many traits that inform consumers about the specific product that may be affected (brand name, lot codes, retail locations where product was sold, etc.). In this respect, maintaining good traceability is protective from an economic standpoint to food companies.

When tracebacks are successful, investigators can also work to identify the "root-cause" of the contamination event that led to adulteration of the food and subsequent human illnesses. When the cause and source of an outbreak are fully understood, these findings can be shared across the affected industry to develop new preventive controls. If an investigation demonstrates negligence on the part of the food company, proper enforcement actions can be taken to ensure that similar violations don't occur again.

The future of improving traceability will rely on leadership from government in defining the minimum requirements for traceability, perhaps building off the CTE and KDE concept. It will also require the food industry to recognize potential economic benefits associated with better recordkeeping such as brand protection and preventing economic adulteration, so that businesses will invest in the processes and technologies needed to collect and maintain the data needed by investigators when a traceback is required. Regulation, technology, and innovation will help to ensure that "fast and right" is no longer a dilemma but rather the standard for tracing food in an outbreak investigation.

References

1. cdc.gov [Internet]. Clifton: Center for Disease Control and Prevention; c2013 [updated 2013 August 6; cited 2015 Sep 17]. Available from: http://www.cdc.gov/pulsenet/next-generation.html
2. Kataoka A, Enache E, Black DG, Elliott PH, Napier CD, Podolak R et al (2014) Survival of Salmonella Tennessee, Salmonella typhimurium DT104, and enterococcus faecium in peanut paste formulations at two different levels of water activity and fat. J Food Prot 77(8):1252–1259. https://doi.org/10.4315/0362-028X.JFP-13-553
3. McEntire J, Bhatt T Pilot projects for improving product tracing along the food supply system-final report [Internet]. Chicago: Institute of Food Technologists; c2013 [cited 2015 Sep 18]. Available from: http://www.fda.gov/downloads/Food/GuidanceRegulation/UCM341810.pdf
4. Miller BD, Rigdon CE, Ball J, Rounds JM, Klos RF, Brennan BM et al (2012 Feb) Use of traceback methods to confirm the source of a multistate *Escherichia coli* O157:H7 outbreak due to in-shell hazelnuts. J Food Prot 75(2):320–327. https://doi.org/10.4315/0362-028X.JFP-11-309
5. Outbreak of Salmonella serotype saintpaul infections associated with multiple raw produce items — United States [Internet]. Atlanta: CDC: c2008 Aug [cited 2015 Sept. 18];57(34); 929–934:[about 4 pages]. Available from: http://www.cdc.gov/mmwr/preview/mmwrhtml/mm5734a1.htm
6. Salmonella Saintpaul outbreak traceback & distribution partial view of the traceback & distribution of peppers from Mexico: July 16 – July 30, 2008 [Internet]. [Location unknown]: HHS/FDA; c2008 [cited 2015 Sept 18]. Available from: http://www.fda.gov/downloads/NewsEvents/PublicHealthFocus/UCM179981.pdf

Chapter 4
Industry Benefits

Michele Southall

Abstract Traceability is a supply chain visibility application that leverages event-based information about products to support track and trace, and recall. However, traceability is *only one application* of that information. As described in the real-world examples provided in this chapter, supply chain visibility information can also be used to enhance and improve *many business processes*, providing companies with on-going incentives and opportunities to improve their business and the bottom line.

Companies implementing supply chain visibility can make use of the data to achieve operational efficiencies and process improvements for inventory management, category management, asset management, quality management, and/or demand forecasting. Improving supply chain visibility can also help reduce errors during procurement and order fulfilment, thereby decreasing error rates and improving selection accuracy. Moreover, supply chain visibility can promote brand reputation and consumer confidence, advance food safety, and strengthen sustainability efforts. Supply chain visibility has bottom line benefits that drive a real-world return on investment (ROI) for implementing or enhancing traceability and supply chain visibility programs.

Keywords Traceability · Visibility · Supply chain · Track · Trace · Recall · Benefits · ROI · Food · Safety

Abbreviations

CEO	Chief Executive Officer
FDA	United States Food and Drug Administration
FSMA	Food Safety Modernization Act

M. Southall (✉)
GS1 US, Chicago, IL, USA
e-mail: msouthall@gs1us.org

J. McEntire, A. W. Kennedy (eds.), *Food Traceability*, Food Microbiology and Food Safety, https://doi.org/10.1007/978-3-030-10902-8_4

GDSN	GS1 Global Data Synchronization Network
GFTC	Global Food Traceability Center
IFT	Institute of Food Technologists
PMA	Produce Marketing Association
POS	point-of-sale
ROI	return on investment
UFPA	United Fresh Produce Association

Supply chain visibility processes, like traceability, are collaborative efforts between trading partners that rely on supply chain standards. As a supply chain standards organization that works with supply chain participants, trade associations, and regulatory agencies across numerous industry sectors, GS1 US[1] has a unique perspective and level of expertise on supply chain visibility, as well as the benefits that it offers.

Introduction

The greatest value of standardized data lies in trading partners' ability to share supply chain visibility information and "see" what is happening along the supply chain [5, 7, 15, 17]. Traceability is a supply chain visibility application that leverages event-based information about products to support track and trace, and recall. However, traceability is *only one application* of that information. Supply chain visibility information can be used to enhance and improve *many business processes*, and these applications provide business reasons to have better supply chain visibility that go beyond public health and safety [5, 6, 10, 16]. For example, the Buyer Operations Study conducted by GS1 US with meat, seafood, and produce suppliers, along with distributors, retailers, and restaurant operators, found that improving supply chain visibility processes results not only in enhanced food safety, but also reduced operational costs leading to incremental increases in revenue generation [4, 9].

Supply chain visibility implementations provide many benefits to many industries (e.g., consumer packaged goods, retail/grocery, foodservice, healthcare, etc.) [15]. The food industry in general, and the fresh category in particular, have a vested interest in enhancing supply chain visibility due to unique industry drivers like the perishable nature of products, volatility of markets, and sensitivity to unexpected

[1] GS1 US, a member of the global information standards organization GS1®, brings industry communities together to solve supply-chain problems through the adoption and implementation of GS1 Standards. Nearly 300,000 businesses in 25 industries rely on GS1 US for trading-partner collaboration and for maximizing the cost effectiveness, speed, visibility, security and sustainability of their business processes. They achieve these benefits through solutions based on GS1 global unique numbering and identification systems, barcodes, Electronic Product Code (EPC®)-enabled RFID, data synchronization, and electronic information exchange. GS1 US also manages the United Nations Standard Products and Services Code® (UNSPSC®). www.GS1US.org

logistics challenges (e.g., weather, transportation hiccups, etc.). Supply chain visibility implementations can advance food safety, strengthen sustainability efforts, and improve business process efficiencies [7, 9, 17] to provide the food industry with numerous benefits, including:

- Operational efficiencies derived from visibility in the supply chain [9, 17]
- Significant process improvements such as labor savings from item scanning with increased self-checkout penetration, better inventory management, more accurate ordering, improved product availability, and improved shrink management [9, 17]
- Reducing the economic impact of food safety emergencies by expeditiously isolating affected products and restoring consumer confidence [9, 17]

Collectively, GS1 US estimates these benefits could represent approximately $3 billion to the fresh foods industry alone, based on a conservative extrapolation of industry findings, pilots, and case studies [9, 17]. Indeed, the food industry is ripe with opportunity to reap real business results from the transformational effects of supply chain visibility [9].

Whole-Chain Visibility

Whole-chain visibility is the combination of internal and external traceability processes, meaning a company's internal data and processes used within their own operations to track a product is integrated into a larger system of external data exchange and business processes that take place between trading partners [4, 9, 17]. To enable that kind of collaboration, supply chain visibility applications are based on the unique identification of products and the standardized exchange of product data at critical tracking events throughout the supply chain. Lessons learned during supply chain visibility implementations (some of which are discussed throughout this chapter) make clear that globally unique identification of products and the standardized exchange of product data at critical tracking events through the supply chain contribute to food safety while also establishing cost-efficient business processes for information linkages to all participants in the supply chain [8, 9, 16, 17].

Benefits All Segments of the Supply Chain

The benefits of supply chain visibility are not limited to only certain segments of the supply chain [4, 9] (see Table 4.1). In fact, opportunities for significant benefits are widespread, [5, 6, 9] as found by the Institute of Food Technologists (IFT).[2] In 2013,

[2]Founded in 1939, the Institute of Food Technologists is committed to advancing the science of food. The non-profit scientific society—more than 17,000 members from more than 95 coun-

Table 4.1 Benefits of supply chain visibility to food industry trading partners [8, 9]

Trading Partner	Benefits
Processor/manufacturer	Enables consumer brand recognition Establishes the foundation for product traceability & recall Optimizes business processes based on accurate product identification for real-time updates and efficient logistics operations Improves visibility into product movement Provides operational efficiencies in receiving, inventory management, and shipping processes Reduces costs and time spent on manual processes Enhances traceability to support safe handling processes Improves order accuracy
Distributor/wholesaler	Optimizes receiving productivity Improves inventory management Increase pick rates Reduces errors in mispicks and shorts
Distribution center (private or 3PL)	Optimizes receiving productivity Improves inventory management Increase pick rates Reduces errors in mispicks and shorts
Point of sale or service	Increases operational efficiencies across various business processes Improves preparedness for fast and precise recalls Provides additional product information to enhance operations for traceability Enables consumer protection

IFT conducted a pilot in coordination with the food industry to examine methods for rapid and effective food traceability, and prepared a detailed pilot report[3] ("IFT Pilot Report") for the United States Food and Drug Administration (FDA) pursuant to the Food Safety Modernization Act (FSMA) [11]. During the pilots, IFT interviewed pilot participants to gain insight into the benefits realized by different segments of the supply chain. In addition to the benefits of traceability, pilot participants across the supply chain described numerous additional benefits they achieved by having supply chain visibility, including:

- Distributors – enhanced inventory management capability, allowing better information for sales forces, improved pick rates, and decreased inventory shrinkage.

tries—brings together food scientists, technologists and related professionals from academia, government, and industry. www.ift.org

[3] Pilot Projects for Improving Product Tracing along the Food Supply System. Institute of Food Technologists (IFT). 2013.

One of the pilot participants estimated a combined cost savings and sales increase of $500,000–$600,000 [1].

- Shippers – real-time supply chain visibility of what products were actually field packed reduced daily overselling and/or underselling, and real-time tracking from field to coolers enhanced the ability to prioritize loads going into the coolers based on when products were picked [1].
- Retailers – better inventory management results, including increased inventory accuracy, selection efficiency, and supply chain visibility [1].
- Growers – improved productivity, accuracy gains in ensuring that the right products are sent to the right customers, and back-office billing efficiencies by being able to more easily determine the recipients of products [1].

Benefits to Inventory Management

Beyond traceability, companies implementing supply chain visibility can also make use of the data to achieve operational efficiencies and process improvements for inventory management, asset management, quality management, and/or demand forecasting [7, 8, 15]. The supply chain efficiencies that can be gained are real and have bottom line benefits [11]. For example, enhancing inventory and category management processes is crucial for perishable products where hours matter in terms of freshness and getting food to consumers at their peak quality [10, 17]. The interoperable, automated processes put in place for supply chain visibility can also help any product category move through the supply chain quickly and efficiently, as demonstrated by Growers Express.[4]

> Growers Express is the supplier to Green Giant Fresh branded lettuces and mixed vegetables who leveraged their traceability program to optimize inventory management. Growers Express stands out in the fresh produce industry because of its aggressive implementation of case-level labeling for 100 per cent of its commodities, which is rare for this sector [6]. What is also notable about Growers Express early and all-inclusive implementation is the reason it did so: the leaders at Growers Express were motivated by the internal benefits to their business that supply chain visibility would deliver [6].
>
> Growers Express manages 25 crews harvesting over 40 varieties of leafy greens and vegetables across several states and international borders, and it rotates its crops up to four times a season to ensure healthy nutrient-rich soil and mitigate diseases. Keeping track of ever-changing crops, harvest dates and inventory movements was challenging. "Because

[4]Growers Express was founded in 1987 by eight produce growers who all believed in a few simple values: producing its own premium quality products, consistent supply and superior service. Growers Express owners have taken three generations of knowledge and respect for the land and have developed it into one of the nation's largest suppliers of fresh vegetables. Headquartered in Salinas Valley, the company's total year-round ground base exceeds 50,000 acres. To offer 40-plus items on a year-round basis, Growers Express also grows in Arizona, Mexico, Oregon, and Ohio. Its largest volume items – iceberg lettuce, broccoli, cauliflower, green onions, celery and leaf lettuces – are complemented with a full line of bunched items. "Green Giant Fresh" chose Growers Express to exclusively license branded lettuces and mixed vegetables. Its vegetables stand up to the stringent quality guidelines for "Green Giant Fresh" trademark. In addition to the "Green Giant Fresh" brand Growers Express also offers its own premium brand "Capurro Farms." www.growersexpress.com

> there is intense pressure in terms of timing, we need to hit on all cylinders at all times and on all crops," explains Jamie Strachan, CEO, Growers Express. "As you can imagine, it's like an air traffic control tower." Their program provided visibility of the date of harvest, the produce harvested, the acreage from which the produce is harvested (including country of origin), and even the precise crew that conducted the harvest – greatly enhancing inventory and supply chain management [6].
>
> "We're always striving for that balance when supplying produce – never lacking what our retailers need versus having too much on hand. With products that have short shelf lives, we have to really manage our inventory, cut-to-cool times, field work order priorities, and inbound logistics. That's key to keeping customers happy and being efficient. Having the necessary real-time, virtual inventory information helps us be more efficient and accurate. When a half-day or a day is gained, finding that 'equilibrium' is greatly enhanced" [6].

Supply chain visibility provides valuable process improvements for inventory and category management processes on the demand-side as well. In terms of inventory management, the unique identification of product by Batch/Lot Numbers and date of receipt facilitates a more efficient "first in, first out" inventory management philosophy, and can be used to establish alerts for stock rotation to maximize quality and minimize spoilage. It also enables retailers to more efficiently facilitate automatic price markdowns as expiration dates grow near, and to prevent expired product from being sold. If products are uniquely identified with additional data detail, point-of-sale (POS) can be encoded for automatic price markdowns, expiration date alerts, even recalled product alerts. In terms of category management, unique identification and additional data on products provides supply chain visibility that can "call out" good brands over "not so good" brands. Retailers with enhanced supply chain visibility have insight to identify when new inventory is required by demographics, and determine what products sell best in what areas [3, 9, 15].

Benefits to Brand Reputation

Moreover, supply chain visibility can promote brand reputation and consumer confidence [8–10, 17]. Visibility information can support decisions that impact brand reputation, thereby improving decision-making [1]. For example, one national restaurant chain mounted an ambitious project to achieve whole supply chain traceability in order to fulfill its brand promise for best ingredients, local sourcing, and sustainable business practices. The company needed to effectively engage with a large network of supplier partners to establish a company-wide traceability process for sharing standardized product information at every step along the supply chain. They found that traceability supports their brand promise by promoting a more transparent supply chain in which they know exactly where the food comes from and can partner with suppliers for continuous improvement in quality, safety and sustainability. In addition, they found that having a traceability system in place simplifies the effort to on-board new suppliers, including more local and regional suppliers, and work with existing supplier partners to create greater supply chain visibility and assurance of the very best sources of food to be found to gain customer recognition.

Benefits to Procurement and Order Fulfilment

Improving supply chain visibility can also help reduce errors during procurement and order fulfilment, thereby decreasing error rates and improving selection accuracy. Unique identification and additional data promotes the right product choice, enables efficient and valid receipt of product, and can lead to electronic proof of purchase/receipt for more efficient order-to-cash processes [8]. In a related effort, IFT worked with the Produce Marketing Association (PMA) and the United Fresh Produce Association (UFPA) to collect information from 18 companies to better illustrate real-world costs and benefits of improving product tracing policies, procedures and technologies [1]. The participating companies identified numerous and significant improvements to their procurement and order fulfilment processes, including:

- One distributor reduced delivery errors from 1/150,000 to 1/250,000, and ultimately expected to reach zero when their system was fully implemented [1].
- A shipper was able to reduce quality claims from over 5% to 1%. These benefits alone paid for entire cost of implementation [1].
- Another shipper achieved 100% accurate order filling/shipment [1].
- Another distributor increased productivity by 15–20%, and decreased selection errors from 1/1500–1/5000 [1].
- A buyer increased order assembly accuracy from 99.5 to 99.99% [1].

Reducing the Financial Impact on Affected Companies

Good supply chain visibility with a solid traceability program enables companies to reduce the economic impact of food safety emergencies by shortening the efforts to remove the product from consumption [4–6, 10]. Recalls cost the fresh foods industry more than $1 billion each year (based on the average cost of the top recalls in fresh food categories over a period of 10 years) [9]. Costs associated with a recall can be staggering for the individual companies involved. For example, the 2008 recall of 143 million pounds of beef cost the Westland/Hallmark meat company about $117 million [9].

A whole-chain traceability process can help companies to efficiently identify and isolate the true cause of an outbreak and react quickly and with precision to avoid a 100% discard of *potentially affected* product [5, 6]. For example, Frontera Produce[5]

[5] Founded in 1992, Frontera Produce is a progressive leader in the fresh produce industry focusing on the changing needs of our customer base. Headquartered in Edinburg, Texas, our diverse growing operations allow us the ease and flexibility of meeting volume, seasonal, regional, and custom packaging requests. After 18 years of evolving our business to exceed our customer's needs, we currently offer a year-round solution of fresh products from all major United States growing areas as well as Mexico, Central and South America. To learn more, visit www.fronteraproduce.com

was able to limit the scope of a cilantro recall to just 12% of total cases – an 88% improvement [5]. A conservative extrapolation to the broader industry (i.e., 25% improvement due to precise recalls instead of Frontera's 88% improvement) would suggest a meaningful opportunity for the broader fresh foods industry to save approximately $250–$275 million a year [9].

In addition, traceability processes enable a company to rapidly pinpoint suspect products anywhere in the supply chain, so they can remove recalled products to keep their customers safe and avoid further risks to consumer confidence [5, 9, 14]. Enhanced supply chain visibility and consumer awareness of the merits and effectiveness of fresh food traceability should also help consumers regain their confidence in the affected product category and return to their previous buying habits in a shorter period of time, helping to bring monthly product sales back to normal levels more quickly [9]. Moreover, traceability processes help companies to reduce unforeseen costs (e.g., legal, fines, forced renovation, lost contracts, loss of customer loyalty) and minimizes collateral damage to supply chain participants and consumers [9].

Reducing the Financial Impact of a Recall on the Affected Product Category

The truth is, even if a company is not linked to a food safety emergency, they are not immune to its devastating financial effects [17]. Health risks associated with contaminated food result in loss of consumer confidence that can damage the entire product category [9]. The fruit and vegetable industries are all familiar with the infamous cases that are still vivid reminders of how a commodity recall can hurt the entire category for months and years if consumers don't have confidence that the industry is able to isolate the affected product quickly and efficiently [17].

For example, the 2006 spinach recall cost the spinach industry $37–$74 million in immediate economic losses, and $350 million in the year following the recall [9, 17]. The contaminated spinach originated from a single 2.8-acre field on one California farm. Nonetheless, it resulted in five deaths and the spread of approximately 200 life-threatening illnesses in 26 states coast-to-coast, an alarming geographic spread of illnesses. The loss in consumer confidence devastated the category during the recall, and made it hard for them to recover: spinach sales were 20% below 2006 levels a year after the recall, and 10% below 2 years afterward [9, 17].

Only 17% of Americans eat spinach [8]. Recalling a product consumed by the majority of Americans would have an even more severe impact, for example:

- A mistaken 2008 *Salmonella* finding in tomatoes cost Florida's tomato industry $500 million [9, 17].
- 2009 *Salmonella* outbreak in peanut butter cost U.S. peanut producers $1 billion [9, 17].

It is sound business strategy to be well prepared for product withdrawals or recalls [17]. Enhanced supply chain visibility and consumer awareness of the merits and effectiveness of traceability can help consumers regain their confidence in the affected product category and return to their previous buying habits in a shorter period of time, helping to bring monthly product sales back to normal levels more quickly. Establishing a comprehensive system for product traceability and food safety optimizes preparedness to address food safety concerns, and helps companies to account for their products and distinguish them from the recall to protect sales and their brand during and after a recall [9].

Return on Investment

The benefits of supply chain visibility that can be achieved across all of the various business processes described throughout this chapter drive return on investment (ROI) for supply chain visibility programs. The associated process improvements and operational efficiencies realized across the organization need to be considered in the ROI analysis for any effort to implement or enhance traceability and supply chain visibility programs.

One tool that can assist with that effort is the Global Food Traceability Center (GFTC)[6] *Seafood Traceability Financial Tool,*[7] which helps organizations in the seafood industry assess the financial impact (costs and benefits) associated with implementing traceability. The *Seafood Traceability Financial Tool* calculates customized traceability ROI based on costs and benefits linked to the optimization and streaming of business activities, expenses that result from implementing traceability tools and systems, and how quickly benefits can be achieved to offset costs to make investment viable [12]. The GFTC has found that opportunities to reap the benefits of traceability can come from:

- New and expanded markets and customers
- Reduction of business liability costs
- Higher recall efficiencies and lower rework costs
- Reduction of product waste and shrinkage
- More reliable and readily accessible data exchange with partners
- More rapid and lower costs of regulatory compliance [12]

These opportunities are not limited to seafood, and companies in other sectors can evaluate all of these benefits in the context of their own sector/operations, and consider them along with the other benefits discussed in this chapter to develop a real-world ROI for implementing or enhancing traceability and supply chain visibility programs [12].

[6] Launched in 2013, the Global Food Traceability Center (GFTC) is a collaborative partnership including public and private stakeholders, created to address the challenges and opportunities of global food traceability implementation. http://www.ift.org/gftc.aspx

[7] https://seafoodtraceability.org/

Keys to Success

Collaboration and a holistic approach are the keys to success for any supply chain visibility program [4, 11], and are essential for achieving the benefits discussed throughout this chapter.

Collaboration helps companies to develop a knowledge base for success. Different supply chain roles, industry drivers/motivations, and even different functions within your own organization all factor into a successful visibility implementation [5]. Collaborating within your organization, with trading partners, and with peers provides key insights. Seeking out the guidance and advice that is born out of that kind of collaboration is what will propel success in developing successful solutions.

A holistic approach means learning, understanding, sharing and applying best practices – including best practices from other industries with similar supply chain visibility/traceability goals [5]. Such an approach enables companies to propel further using solutions that are tested, stretched, and validated. It enables companies to avoid "reinvention," and possibly identify a peer who has the same issue but may already have a great solution figured out.

Collaboration and a holistic approach are best led by industry initiatives with neutral leadership through trade associations[8]; standards bodies,[9] or other collaborative forums.[10] Those who are most concerned and knowledgeable often seek out opportunities to contribute to the collective knowledge, and these organizations provide excellent avenues for such contributions. Moreover, these types of industry initiatives provide a forum for collaboration that can bring different supply chain roles together to develop solutions for industry issues and opportunities.

This chapter presented examples of real value achieved in both initial implementations, as well as implementations to enhance existing supply chain visibility programs. As these examples illustrate, on-going attention to supply chain visibility is essential to stay on track and keep improving, and to avoid letting the knowledge base slip. Supply chain visibility can be leveraged to improve numerous business processes, providing companies with on-going incentives and opportunities to improve their business and the bottom line.

Looking Ahead

Industries, markets, and companies are dynamic and ever-evolving. Beyond supply chain visibility, the next frontier is data quality. In fact, industry leaders are building on the progress achieved with their supply chain visibility programs by pursuing data quality.

---

[8] e.g., Grocery Manufacturers Association (GMA); Produce Marketing Association (PMA); etc.

[9] e.g., GS1 US; International Organization for Standardization (ISO); American National Standards Institute (ANSI); etc.

[10] e.g., Produce Traceability Initiative (PTI); National Fisheries Institute (NFI); etc.

For example, Dot Foods (Dot),[11] a national food redistributor is using quality data as a key enabler for growth. Dot introduced its online product catalogue, the Dot Expressway® ("Expressway"), 18 years ago to provide better search capabilities for its customers and sales representatives. Dot launched a data quality initiative that transformed the Expressway to include expanded and validated data attributes. Dot suppliers publish core item attributes such as case dimensions and weights as well as marketing and nutritional information like ingredients, allergens and images through data pools that are then synchronized with Dot's system via the Global Data Synchronization Network™ (GDSN®). With accurate data, Dot has reduced costs with improved warehouse management and load optimization, and increased productivity [2].

Another industry leader in the area of data quality is Shamrock Foods[12] (Shamrock). Shamrock needed complete and accurate data for its web-based ordering system to satisfy the increasing demands of its customers, especially healthcare providers and school systems that require nutritional and allergen information before purchasing. The company also understood how using quality data could help it become more efficient and cut costs throughout its operations. The Shamrock team created a new supplier connection process for its vendors to publish their product data and images in its electronic ordering system for customers. Today, more than 500 Shamrock suppliers publish detailed product information as well as share and synchronize data via the GDSN. Shamrock has experienced a 20% increase in sales and a 14% increase in demand for those products that have extended data attributes such as nutritional information and product images. The company has increased its level of customer service by providing extended product information to help them make better buying decisions [13].

The benefits of supply chain visibility described in this chapter and the benefits of data quality achieved by these industry leaders demonstrate how advancements in technologies and standards have created a ripe environment for improving business processes and operational efficiencies [11]. The opportunities are vast and implementation stories documenting how companies are overcoming challenges are plenty.

[11] Dot Foods Inc. is the nation's first and largest food industry redistributor, serving all 50 states and 25 countries from nine distribution centers. For more than 50 years, Dot has developed innovative solutions that not only benefit its own operations, but also those of its manufacturers, distributors and distributors' customers—foodservice operators ranging from small restaurants to large institutions.

[12] Founded in 1922 and still family-owned, Shamrock Foods Company specializes in the manufacturing and distribution of quality food and food-related products. The company has become a symbol of integrity in households and businesses. Shamrock continues serving customers through a family of companies, including Shamrock Farms, the largest dairy in the Southwest, and Shamrock Foods, the seventh-largest U.S. foodservice distributor.

References

1. Carpenter B GS1 US outlines roles in key initiatives [Internet]. [Location unknown]: Supermarket News. 2014 [cited 2015 Nov 3]. Available from: http://supermarketnews.com/technology/gs1-us-outlines-roles-key-initiatives
2. Dot Foods – National food redistributor uses quality data as expressway for growth [Internet]. Lawrenceville: GS1 US; 2015 [cited 2015 Jun 9]. https://www.gs1us.org/documents?Command=Core_Download&EntryId=1574
3. Fernandez A Better food traceability to reduce unnecessary waste. Food Manufacturing [Internet]. 2014 [cited 2015 Apr 27]. Available from: http://www.foodmanufacturing.com/articles/2014/10/better-food-traceability-reduce-unnecessary-waste?et_cid=4197576&et_rid=703187341&type=cta
4. Fernandez A. Traceability update: regulatory and cultural trends shift from response to prevention. Food Safety Magazine [Internet]. 2015 [cited 2015 Apr 27]. Available from: http://www.foodsafetymagazine.com/magazine-archive1/februarymarch-2015/traceability-update-regulatory-and-cultural-trends-shift-from-response-to-prevention
5. Frontera Produce: traceability from field to store [Internet]. Lawrenceville: GS1 US; 2010 [cited 2015 Apr 27]. Available from: https://www.gs1us.org/documents?Command=Core_Download&EntryId=512
6. Growers Express: express route to traceability [Internet]. Lawrenceville: GS1 US. GS1 US; 2011 [cited 2015 Apr 4]. Available from: https://www.gs1us.org/documents?Command=Core_Download&EntryId=513
7. GS1 US seafood traceability readiness program proof of concept [Internet]. [Location unknown]: National Fisheries Institute and GS1 US; 2014 Nov [cited 2015 Sep 5]. Available from: http://www.gs1us.org/gs1-us-library?Command=Core_Download&EntryId=1327&method=attachment
8. Improving traceability and food safety with GS1 standards in fresh foods [Internet]. Lawrenceville: GS1 US; 2012 [cited 2015 Apr 27]. Available from: https://www.gs1us.org/https://www.gs1us.org/documents?Command=Core_Download&EntryId=553
9. Integrated traceability in fresh foods: ripe opportunity for real results [Internet]. Lawrenceville: GS1 US; 2013 [cited 2015 Apr 13]. https://www.gs1us.org/documents?Command=Core_Download&EntryId=598
10. JemD Farms: Greenhouse grower leads the way with GS1 standards and the produce traceability initiative to achieve whole chain traceability [Internet]. Lawrenceville: GS1 US; 2012 [cited 2015 Apr 13]. https://www.gs1us.org/documents?Command=Core_Download&EntryId=785
11. McEntire J, Bhatt T Pilot projects for improving product tracing along the food supply system [Internet]. Chicago: Institute of Food Technologists. 2012 [cited 2015 Apr 27]. Available from: http://www.fda.gov/downloads/Food/GuidanceRegulation/UCM341810.pdf. Accessed 13 Apr 2015
12. Seafood traceability financial tool [Internet]. Chicago: Global Food Traceability Center. 2014 [cited 2015 Jun 27]. Available from: https://seafoodtraceability.org/
13. Shamrock foods – foodservice leader serves up quality data from pallet to palate [Internet]. Lawrenceville: GS1 US; 2015 [cited 2015 Jun 9]. https://www.gs1us.org/documents?Command=Core_Download&EntryId=1572
14. Smith T. SmartBlog on food & beverage [Internet]. [Location unknown]: SmartBrief, SmartBlogs; 2014 – Q-and-A: GS1US' Angela Fernandez on the importance of supply chain traceability; [cited 2015 Apr 27]. http://smartblogs.com/food-and-beverage/2014/02/12/q-and-a-gs1us-angela-fernandez-on-the-importance-of-supply-chain-traceability/
15. The GS1 US visibility framework [Internet]. Lawrenceville: GS1 US; 2012 [cited 2015 March 3]. Available from: https://www.gs1us.org/documents?Command=Core_Download&EntryId=1473
16. The Oppenheimer group: global produce distributor uses the advance ship notice for efficiently sharing traceability data [Internet]. Lawrenceville: GS1 US; 2014 [cited 2015 Jun 8]. Available from: https://www.gs1us.org/documents?Command=Core_Download&EntryId=530
17. Traceability and food safety Q&A with Angela Fernandez: whole-chain traceability could save billions of dollars. Fruit Growers News. 2014 [2015 Sep 5]

Chapter 5
The Traceability of Bulk Food Products

Edward Orzechowski

Abstract Bulk food products and ingredients serve as the building blocks for the hundreds of thousands of food products sold today by retailers, restaurants and foodservice globally. Bulk food products are also used as feed for livestock. Therefore, heightened awareness and proactive food safety procedures are required since the impact of a contaminated ingredient can be widespread. Despite the implementation of good practices in bulk material processing and distribution, it can be expected that issues may occasionally arise which will trigger the need to trace the supply chain pathways of bulk materials. This chapter is focused on the need and drivers responsible for the implementation of traceability systems in the bulk supply chain for both liquid and dry products. The definitions and defined stakeholders in this chapter are unique and specific to the challenges associated with tracing bulk food products. The motivation for traceability must be viewed more broadly than in the context of a safety procedure but also as an integrated concept with respect to sustainability, technology, and cost optimization.

Keywords Grains · Commodities · Oils · Powders · Crystals

Introduction

The traceability of bulk food products, both backwards or forwards, pose a unique challenge within the food industry. Whether in dry or liquid form, bulk products tend to be treated similarly to fluids when traced throughout the food system because of the nature of the products. The supply chains of bulk food products are more accurately described as a food system because of the intricate web that is created by bulk products. Their path from the farm or supplier, all the way to their end point is not a linear path. The product is often blended with the same products from different lots or batches (such as in silos or tankers), or with other products entirely (such as in the manufacturing of processed foods). As discussed throughout other chapters,

E. Orzechowski (✉)
ASI Government, Washington, DC, USA

J. McEntire, A. W. Kennedy (eds.), *Food Traceability*, Food Microbiology and Food Safety, https://doi.org/10.1007/978-3-030-10902-8_5

taking a holistic approach can help improve food security, recall management, risk mitigation, foodborne illnesses, freshness, food waste and transparency.

The simplest path of bulk products starts as raw ingredients which are made into partial or mixed products that then become finished products. In this chapter, bulk food products are defined as raw or partially processed food ingredients or products that play a role in a given food system. The chapter does not attempt to define a particular volume or weight as "bulk". Rather, one attribute that influences traceability is the lack of "clean breaks", particularly during storage (such as a silo that is continuously filled with newly produced bulk product).

Bulk food products are particularly susceptible to incidental cross-contamination or mixing from comingling in storage. This is due to both the natural composition of these products and the cost and practicality of preventing comingling. Raw materials are harvested and the ingredients or products are typically stored in containers such as tanks, bins, silos and other appropriate vessels. During production, raw or partially processed products can be processed by chemical and/or microbiological means such as pasteurization, sterilization, concentration, dilution, and fermentation. These changes to products then requires the generation of production lot / batch codes. Tracking this information becomes critical when the production process is not continuous and the intermediate product must then be stored. Even when the processing steps are well defined (e.g. "batch" production) with easily distinguishable lot numbers, this differentiation can be difficult to maintain during storage.

Traceability for bulk food products is defined as the ability to determine each portion of a raw, intermediate, or final product's composition in terms of unit inputs and outputs to the system. Historically the need to identify these proportions of products has been driven by food recall management but as the U.S. food system has evolved, transparency has emerged as a key driver for food companies. The value of maintaining a full traceability program across all product lines has shifted from defensive measures towards opportunistic measures for efficiency in food companies.

The following are four types of products that are typically classified as bulk food products:

1. Liquids – Fluids that are fully or partially free flowing and typically create a mixture when combined with other substances. These include milk, vegetable oils, juices, and other viscous compounds.
2. Grains – Either cereal crops, or seeds or kernels of cereal crops, that are typically stored in silos or bins and have historically been traced to the elevator of a grain processing facility. These include wheat, rice, corn, oats and other grains as well as those used in animal feeds.
3. Powders and Crystals – Solids that are reduced to a fine and loose-particle substance that can be derived from both agricultural and non-agricultural sources. These include spices, dry ingredients, and vitamins/minerals. They also include solids that have a crystalline structure that include sugars, salt and other similar ingredients.

4. Bulk Commodities – Food products that have loose forms that are typically measured through their value chain by their weight. These types of products can include coffee, cocoa, tea and similar products.

For a traceability system to be functional and work correctly, it must be collectively exhaustive. There cannot be gaps, or the tracing forward or backward will be flawed. A food system that involves bulk products must have all inputs and outputs tracked. Mass balance must be maintained; what comes into the system must come out with a minimal margin of error to account for natural loss. For bulk products to be completely tracked through the food system, at each point of the harvesting, manufacturing, collection points and transportation process there needs to be a record of all the relevant information including but not limited to the batch or lot number (which may be a harvest or production date), weights, handling conditions, and processing information.

A simplified method for tracking foods through the manufacturing process is using cleaning-in-place procedures for the containers of bulk products. This is a feasible system for manufacturers who can take the time to completely clean out silos, bins, tanks and other storage containers in between batches. This is a cost-effective system for some powders and crystals that may be stored in smaller/disposable containers (i.e. bags and plastic bins) but are not for products such as liquids and grains which typically use larger containers within the manufacturing process. This is also inefficient for manufacturers who use continuous flow processing.

When manufacturers use continuous batch processing, they typically create batches that contain products over a period of time instead of a specific end-to-end lot. While this allows manufacturers to have a system in place to track what the possible input lots were that went into a series of finished (or intermediate) batches, it makes precision tracking much more difficult. For example, if you are making a bakery product like cookie dough in a continuous flow process, each ingredient is added at different stages of the production process at their varying proportions by weight. Each of those ingredients is unlikely to be depleted at the same rate and new ingredient lots are likely needed to be used to keep production going. Depending on the amount of time used for each period, the amount of ingredient used in relation to its container and the nature of the product, a single ingredient lot could span many production lots and vice versa. With smaller time periods, there is a higher chance of single ingredient lots spanning multiple finished product lots, while with larger time periods, it is more likely that single finished product lots contain multiple ingredient lots. The nature of food production leads manufacturers to make this decision based on operational efficiency. Generally, longer periods between batches translates to lower operating costs for manufacturers. In these situations, traceability may be complicated, but is not impossible. The inputs and outputs should still be tracked. In other words, in the event of a food safety or quality issue, while the amount of potentially affected product may be large, the scope of the issue is much more readily ascertained when lot and batch numbers are captured as opposed to assuming traceability of bulk products is futile, and not capturing any information at all.

Definitions

While traceability has been defined and discussed in general terms, it is important to highlight how definitions pertain to the nature of bulk traceability.

Traceability was first defined in ISO 8402 as: "The ability to trace the history, application or location of an entity by means of recorded identifications." This definition clearly states what should be traced (history, application and location) and also how the tracing should be done (by means of recorded identifications) [9]. This definition was withdrawn by ISO and superseded by ISO 9000. As per the definition in ISO 9000 traceability is defined as "Ability to trace the history, application or location of that which is under consideration" [5].

The ISO guidelines further specify that traceability may refer to the origin of the materials and parts, the processing history, and the distribution and location of the product after delivery. This definition of traceability is quite broad. It does not specify a standard measurement for "that which is under consideration" (a grain of corn or a truckload of corn), a standard location size (field, farm, or county), a list of processes that must be identified (pesticide applications or animal welfare), where the information is recorded (paper or electronic record, box, container or product itself), or a bookkeeping technology (pen and paper or computer) [4]. The guidelines do not specify that corn flakes be traceable to the processor/manufacturer, or to the corn field, or to the seed company from which the corn grew.

Drivers & Stakeholders

There have been several studies focusing on traceability and its importance in the food supply chain. The main focus of these traceability studies has been primarily on perishable food products such as meat and produce [1, 6]. With the advent of marketing opportunities, awareness of consumers, and need of new solutions traceability has become more focused. Traceability should be seen as a connection between consumers and stakeholders at each level of the supply chain. In 2004 the USDA listed bulk products as a key area of future focus on traceability [13]. This study identified the traceability of bulk food products such as grain to have been limited due to cost constraints to trace grain to the elevator of the grain processing facility. As consumers have pushed for increased transparency in the market, companies have been driven to change this practice. This positions food companies to hold suppliers and growers to higher standards of tracking their food products.

In the past, some food safety incidents led to increased recognition of the importance of traceability in the food supply. The European Union has prioritized the necessity of traceability by attaching this concept to labelling and consumer concern about mad cow disease, dioxin in chicken feed, and the GM food products [8].

The drive for transparency within the food system and the rise of consumer interest in the contents of their food products has laid the course for improved bulk product tracing systems. This shift in consumer perception has given way to various certification programs to verify the absence of specific products or cross-contamination within processing lines. While there is a new surge in interest for food companies, the concept of preventing cross-contamination of specific food products has been around for decades. For example, keeping wheat flour and other bakery ingredients separate and tracking them uniquely for kosher products is not a new concept. The scale in which this tracing is required is however a new frontier, because of the documentation that accompanies traceability.

Within the past two decades, the debate revolving around genetically modified ingredients has spurred controversy and has driven greater transparency within the food system. A large driver behind this movement has been consumers seeking information about food such as origin, harvest date and whether or not a product contained genetically modified ingredients [8]. This is relevant for both human and animal food products; a major portion of grains, specifically corn and soybeans, are used as animal feed. Bulk products may also have human rights and labor abuse factors driving the establishment of certain traceability programs.

Bulk Traceability Handling

The very basic process for moving a single product from entry to exit has many touch points that require data collection at each step. In the case of bulk products, these intermediate steps can be more challenging to track. This is because as products are brought together or apart, the manufacturer must then create new identifying information such as a lot number, batch number or other unique identifier to trace the products and their components backward and forward. These intermediate steps depend on a manufacturer's approach to handling and processing of their products.

There are two traceability frameworks used to substantiate claims of sustainability or other production practices for food products at each step of the supply chain. These schemes are applied depending on varying levels of controls around the handling of the products. These certification handling frameworks are specifically used on bulk commodity handling in various capacities and as such, have different applications across the different bulk product categories [2]. The two schemes are as follows [12]:

1. Product Segregation – separation of certified and non-certified products physically at each stage of the supply chain. With proper implementation, these systems yield a 100% separation between segregated product groups. This type of model is split into two categories:
   (a) Bulk Commodity – Allows mixing of similar products that are certified from different lots but must stay separate from non-certified products.
   (b) Identity Preservation – Requires products to be segregated from non-certified products and to be kept from mixing with other certified products in the chain.

2. Mass Balance – This model allows certified and non-certified products to be mixed within the value chain. This is permitted as long as the ratio between certified to non-certified products is above a certain threshold. (i.e. 80% certified vs. 20% non-certified) [13].

The application of these different handling methods depends on the nature of the products being traced and other drivers such as cost and best practices. Cost is a driving factor for most suppliers and businesses within the value chain, especially when bulk systems are very complex. The more stringent the controls required for a product to be traced through the system, the more complex means of traceability, which drives up costs. For complex food products, a mixture of the above handling practices can be utilized. Products such as bulk commodities, powders and grains typically follow the Bulk Commodity type of handling where liquids typically fall under the Mass Balance framework. There are exceptions between each and how they are used on a case-by-case basis. Other frameworks and modeling methods have been developed that quantify and measure these traceability systems. They can be found in implementation guides and in traceability research [10].

For both liquids and some free-flowing solids, process controls are used to track the contents within bulk containers like silos and tanks, and as they move from one to another. The inputs and outputs of bulk products are measured by their flow rates. Certain engineering models can use the flow rate along with time, tank dimensions, particle size, viscosity and other factors to calculate the amount of one batch that is still within a container relative to the total in the container. Sensors and monitors can be used to track these factors such as container levels, temperature, and pressure. Most powders and some of the other solid bulk products rely on other tracking systems to track their contents. They can have unique issues in maintaining accurate measurements of content levels in storage containers because they do not flow freely like liquids and some solids. This can cause the material to clump or cake up against the side, dispense at irregular angles and compact over time. Solid bulk products such as grains and commodities can also be greatly affected by their seasonality because the density of the material can vary from season to season. These all make it difficult to accurately know how much of a product has left a container and what is still left, which can reduce the accuracy of some tracing frameworks, models, calculations, and assumptions.

Technology

In recent years, food traceability researchers have expanded their focus beyond food safety to include operational efficiencies – product freshness, inventory availability and waste reduction, for example. Enabling technologies for faster, more precise and less expensive product identification of bulk food products include RFID devices, network enabled sensors, linear and two-dimensional barcodes and advanced diagnostics systems. These tools paired with web connected mobile computing devices can help quickly identify, capture and share information about product lots and batches (albeit for discreet amounts of product) [7]. For example, Internet connected sensors are now widely used as part of integrated moisture monitoring and traceability systems for bulk powder and commodity food products [3]. These monitors allow for better food safety and quality control and, depending on application, can have the side benefit of improved traceability of the entire system. However, challenges still exist between what information is collected and shared throughout the system. Often agricultural commodities have long, complex supply chains that span large geographic areas, cross multiple borders and span companies of various sizes and technological sophistication. The resulting maze of regulatory requirements, incomplete internet coverage and disparity in technology investment creates a challenging environment for maintaining bulk food traceability. In the future, one can expect the cost, effectiveness and availability of food traceability technologies will help solve the technological challenges. However, it will be up to the food industry to work with regulators, buyers and third-party certification schemes to align traceability requirements.

The Future of Bulk Product Traceability

In the past, bulk traceability has largely been driven by the needs of grain and commodity handling practices. The food system, not only in the U.S., but globally, is rapidly evolving to drive all the stakeholders in the food system's value chain to adapt and evolve as well. As consumer needs and wants change, there will need to be a more accurate and uniform means of tracking throughout the system. While making improvements on the operational side of traceability, the future improvements to bulk products lie in the rapid development of technologies that communicate and store information. The integration of emerging technology will aid in the growing need to holistically trace food products throughout the entire system. While this changing landscape brings with it many challenges, there are opportunities for growth in both the public and private sector [11].

References

1. Becker T (2000) Consumer perception of fresh meat quality: a framework for analysis. Br Food J 102(3):158–176. https://doi.org/10.1108/00070700010371707
2. Comba L, Belforte G, Dabbene F, Gay P (2013) Methods for traceability in food production processes involving bulk products. Biosyst Eng 116(1):51–63
3. Fraiture M, Herman P, De Loose M, Debode F, Roosens N (2017) How can we better detect unauthorized GMOs in food and feed chains? Trends Biotechnol 35(6):508–517
4. Golan E, Krissoff B, Kuchler F, Calvin L, Nelson K, Price G (2018) Traceability in the U.S. food supply: economic theory and industry studies. [online] Economic Research Service, U.S. Department of Agriculture, Agricultural Economic. Available at: https://ageconsearch.umn.edu/bitstream/33939/1/ae040830.pdf. Accessed 27 July 2018

5. International Standards Office (1987) ISO 9000 family – quality management. ISO, Geneva
6. Kehagia O, Chrysochou P, Chryssochoidis G, Krystallis A, Linardakis M (2007) European consumers' perceptions, definitions and expectations of traceability and the importance of labels, and the differences in these perceptions by product type. Sociol Rural 47(4):400–416
7. Kvarnström B, Bergquist B, Vännman K (2011) RFID to improve traceability in continuous granular flows—an experimental case study. Qual Eng 23(4):343–357
8. McGarry Wolf M, Bertolini P, Shikama I, Berger A (2015) A comparison of attitudes toward food and biotechnology in the U.S., Japan, and Italy. Food Distrib Res Soc 43(1):103–110
9. Olsen P, Borit M (2013) How to define traceability. Trends Food Sci Food Saf 29:147–150
10. Thakur M (2010) Operational techniques for implementing traceability in bulk product supply chains. Graduate. University of Iowa
11. Thakur M, Mosher G, Brown B, Bennett G, Shepherd H, Hurburgh C (2009) Food safety series—traceability in the bulk grain supply chain. Resour Mag [online] 3:20–22. Available at: https://lib.dr.iastate.edu/abe_eng_pubs/396/. Accessed 27 July 2018
12. United Nations Global Compact & BSR (2014) A guide to traceability a practical approach to advance sustainability in global supply chains. [online] Available at: https://www.unglobalcompact.org/docs/issues_doc/supply_chain/Traceability/Guide_to_Traceability.pdf. Accessed 27 July 2018
13. USDA Advisory Committee on Biotechnology and 21st Century Agriculture (2004) Preparing for the future

Chapter 6
Meat and Poultry Traceability – Its History and Continuing Challenges

Hilary S. Thesmar and Shawn K. Stevens

Abstract When it comes to traceability of food products, some of the most significant challenges and opportunities are found in the meat and poultry industries. This is because traceability becomes substantially more complicated for any product that is derived from a live animal. On the farm, traceability of the animals themselves is critical to avoid the introduction and spread of animal disease. After harvest, traceability can become even more complex as components from hundreds of animals are commingled to process individual food products. This complexity is enhanced at retail, where some of these products are processed further. To manage these complexities, traceability systems have been developed to help better track the origins of raw animal foods. In addition, to enhance their ability to regulate food and respond more rapidly to foodborne illness outbreaks, FSIS has continually tightened the regulations governing the traceability of food products and the animals from which they are derived.

Keywords Meat and poultry traceability · Animal traceability · Ground beef traceability · Traceability systems · Critical tracking events · Key data elements · Traceability technologies · Foodborne illness outbreak investigations

Animal Traceability

Meat and poultry traceability starts with the animal. Animal identification started with farmers marking animals for the purpose of identifying ownership. In more recent decades, animal identification has become important for the purpose of tracing animal disease outbreaks and illnesses [6]. Other reasons to trace animals include added value animals and meat products for marketing purposes, compliance

H. S. Thesmar (✉)
Food Marketing Institute, Arlington, VA, USA
e-mail: hthesmar@fmi.org

S. K. Stevens
Food Industry Counsel, LLC, Milwaukee, WI, USA
e-mail: stevens@foodindustrycounsel.com

J. McEntire, A. W. Kennedy (eds.), *Food Traceability*, Food Microbiology and Food Safety, https://doi.org/10.1007/978-3-030-10902-8_6

with regulations such as country of origin labeling, and biosecurity controls that require proper segregation and separation of animals. Most strong management programs for farms and for food production and manufacturing include traceability programs.

Over the years there have been several large-scale crisis situations that have reinforced our need for animal traceability on an international scale. Foot and mouth disease outbreaks are caused by a highly-contagious virus and result in huge economic losses for farmers and ranchers. The most recent outbreaks have been in the United Kingdom between 2001 to 2007 and were devastating to the livestock industry. BSE is another high-profile animal disease reinforcing the need for animal traceability. Over the past several decades, the need for increased tracing of animals throughout their lives has been reinforced with ongoing suspected and confirmed cases of BSE in cattle. A slow moving but devastating illness, traceability is the key to identifying the BSE exposure points in cattle and minimizing the risk to the industry.

Most countries have some type of animal traceability requirements. In the United States, the United States Department of Agriculture ("USDA") Animal and Plant Health Inspection Service ("APHIS") has the authority to regulate the international movement and the interstate movement of livestock. The current regulations were finalized on January 9, 2013 and require that livestock be identified and that interstate movement be accompanied by an interstate certificate of veterinary inspection or other documentation unless specifically exempted. The purpose of the rule was to help identify diseased animals or at-risk animals in the event of animal disease outbreaks for containment purposes [7].

One crisis that transcended the challenges of traceability from animal to food was the horse-meat scandal in Europe in 2013. Economic adulteration of beef appeared to be the motivation when horsemeat was found in the beef supply in 2013. The initial finding appeared to be limited but, upon investigation, the crisis was a widespread event with many countries in Europe impacted with adulterated beef. Tracing the meat and eventually the animals turned out to be a huge challenge. It is not known what proper tracing systems could have done to avert or identify the problem much sooner. The crisis was economically devastating to the industry and, more importantly, decreased consumer confidence in the food supply. The crisis drew international attention and many questions were raised worldwide.

Industry Initiatives to Develop Traceability Programs

To address data standards within the meat and poultry industries, the Meat and Poultry Business-to-Business Data Standards Organization known as "mpXML" was founded in 2001 to bring multiple partners and stakeholders together to develop a set of cohesive and acceptable standards for the industry to utilize. Multiple standards documents were published over the years and, in 2014, the group merged with the GS1 US Meat and Poultry Workgroup.

GS1 US has since taken over the work of mpXML and the standards are now GS1 standards. The goal of the mpXML organization was to provide standards for identifying trading partners, trading locations, products used or created, logistics units received or shipped, and tracking inbound and outbound shipments. There was a hierarchy of information among products in the supply chain including shipments, case, pallet and consumer packaging. Different information was required for each product according to the hierarchy with consumer packaging requiring human readable information in addition to UPC-A or UPC-Type 2 bar codes [5]. The standards are now published by GS1 US as the Traceability for Meat & Poultry U.S. Implementation Guide [2].

The IFT Global Food Traceability Center, a public-private partnership founded in 2013, has brought stakeholders together to discuss traceability issues and has published a series of reports and papers on traceability throughout multiple food product categories.

Just as is the case for most other foods, critical tracking events (CTEs) and Key data elements (KTEs) are the building blocks of traceability for meat and poultry [4]. To define the terms more clearly:

- Critical Tracking Events ("CTEs"): CTEs are the dots that need to be connected such as linking incoming and outgoing products, transformation (internal traceability), location changes (external traceability), shipping and receiving. The CTEs are the transactions that occur along the supply chain.
- Key Data Elements ("KDEs"): KDEs are the units of data that need to be tracked in order to have complete information throughout the supply chain. For example, lot numbers, quantities, locations, times/dates, and ingredients.

In 2014 as part of the work of the Global Food Traceability Center, Zhang and Bhatt published a comprehensive guidance document on the best practices in food traceability. In the document, each food sector was isolated and specific information provided regarding the traceability issues and the CTEs and KDEs. A comprehensive list of CTEs and KDEs was developed by a group with subject matter expertise and reviewed by many stakeholders and experts in the industry. The list of CTEs and KDEs identified for the meat and poultry supply chain from the guidance document are below [8] (Tables 6.1, 6.2, 6.3, 6.4, 6.5, and 6.6).

The supply chains for meat and poultry products are quite variable. The poultry industry is highly vertically integrated with a single company typically owning the poultry from the hatchery through the processing plant. Contract growers raise the birds and are paid based upon live weight. The beef industry is not vertically integrated with ownership changing multiple times. The pork industry is a mixture of the two models, with some vertical integration.

Traceability becomes a challenge during harvest or any sort of transformation. Maintaining the identity of the former product as it becomes a new product has always been a challenge. The meat and poultry industries are no exception. The practice of comingling adds to the confusion and the risk with additional losses in traceability and product integrity.

Table 6.1 Poultry CTEs [8]

Poultry CTEs				
1	2	3	4	5
Egg delivery	Eggs to incubator	Hatched eggs	Unhatched eggs	Chick delivery to farm
6	7	8	9	10
Chick placement	Shipment of feed to farm	Delivery of feed to farm	Mature broilers/ spent hens	Broiler pickup
11	12	13	14	15
Broiler delivery	Broiler dead on arrival (DOA)	Broiler harvest	Minimally processed meat	Shipping to partner
16	17	18	19	20
Receiving by partner	Nonmeat ingredient	Packaged finished product	Shipping to distributor	Receiving by distributor
21	22	23	24	25
Shipping to retailer or food service operator	Receiving by retailer or food service operator	Retail POS	Case opened by food service operator	Product disposed as unusable waste

Table 6.2 Specialized framework for poultry CTE and KDE [8]

Poultry KDEs					
Who	Where	When	What	Identifiers	Activity Types
Owner of breeder farm		Date	Eggs	Breeding stock	Purchase orders
Owner of hatchery	Location of hatchery	Time	Chicks	Flock ID	Delivery identification
Owner of broiler farm	Location of broiler farm		Feed	Product	Process identification
Owner of feed mill	Location of feed mill		Broilers/spent hens	Batch number/ lot number	Cycle identification
Owner of processing plant	Location of processing plant		Nonmeat ingredients	Use-by date	Feed order number
Owner of cold storage	Location of cold storage		Packaging	Sell-by date	Ticket number
Owner of retail	Location of retail DC/store		Processed product		Work order number
Owner of food service operation	Location of food service DC/ Restaurant				Carrier name
Owner of breeder farm					Trailer number

Table 6.3 Beef CTEs [8]

Beef CTEs				
1	2	3	4	5
Feed	Shipping to processing plant	Receiving by processing plant	Live animals	Minimally processed meat
6	7	8	9	10
Nonmeat ingredients	Packaged finished product	Shipping to distributor	Receiving by distributor	Shipping to retailer/food service operator
11	12	13	14	
Receiving by retailer/food service operator	Retail POS	Case opened by food service operator	Product disposal as unusable waste	

Table 6.4 Specialized framework for beef CTE and KDE [8]

Beef KDEs					
Who	Where	When	What	Identifiers	Activity Types
Owner of feed lot	Location of feed lot	Date	Cattle	Animal identification	Purchase order
Owner of processing plant	Location of processing plant	Time	Feed	Animal batch	BOL
Owner of cold storage	Location of cold storage		Nonmeat ingredients	Product	Feed order
Owner of distributor	Location of distributor		Packaging	Batch number/ lot number	Cycle identification
Owner of retailer store	Location of retail distribution center (DC)/store		Processed product	Use-by date	Ticket number
Owner of food service operation	Location of food service distribution center (DC) / restaurant			Sell-by date	Work order number
					Carrier name
					Trailer number

Table 6.5 Pork CTEs [8]

Pork CTEs				
1	2	3	4	5
Feed	Hogs	Shipping to processing plant	Receiving by processing plant	Minimally processed meat
6	7	8	9	10
Nonmeat ingredients	Packaged finished product	Shipping to distributor	Receiving by distributor	Shipping to retailer/food service operator
11	12	13	14	
Receiving by retailer/food service operator	Retail POS	Case opened by food service operator	Product disposal as unusable waste	

Table 6.6 Specialized framework for pork CTE and KDE [8]

Pork KDEs					
Who	Where	When	What	Identifiers	Activity Types
Owner of finishing house	Location of finishing house	Date	Hogs	Product	Purchase order
Owner of processing plant	Location of processing plant	Time	Feed	Batch number/lot number	BOL
Owner of cold storage	Location of cold storage		Nonmeat ingredients	Animal identifier	Feed order
Owner of distributor	Location of distributor		Packaging	Use-by date	Cycle identifier
Owner of retailer store	Location of retailer		Processed product	Sell-by date	Ticket identifier
				Product	Work order
					Production date
					Trailer number
					Carrier name

Traceability Technologies

Radio-frequency identification (RFID) is a technology that has been available for a few decades but has yet to be widely used for traceability purposes due to cost. In the meat industry, RFID technology is often implanted in cattle ear tags to track the identity and location of cattle. The RFID device is needed as well as a reader to extract the information from the device. Software and a traceability system must also be set up to manage the information. RFID is very useful in that it contains the information in one embedded device with little human interaction [3]. In addition to using the technology to track live animals outside of the processing facility, the technology also can be used in the production environment for enhanced real time tracking and inventory data. The information must be maintained in a software system in order to be useful to the owner or user of the information. A barcode may contain the same information and is much less expensive than a RFID. At this time, RFID is more appropriate for larger volume or higher dollar value items. It is possible that, over time, RFID costs will decrease and become more readily available to the industry.

Additional cutting-edge technologies are also being developed to enhance the traceability of raw animal products even further. New DNA technologies, for instance, can be employed to provide meat packers, processors, grocery retailers and the food service sector with the ability to trace the origin of meat products from point of sale to the animal of origin. By way of example, the traceability of ground beef could be enhanced significantly by taking a sample of DNA from each animal carcass before it is fabricated into trim that will be further processed into ground beef. The characteristics of each sample would be entered into a database. Later, the finished ground beef product could be sampled to determine the specific animals that were used to make that finished product. These types of traceability technologies can also have other benefits as well, allowing supply-chain participants to authenticate and validate meat product attributes such as natural, organic, or Angus.

These types of DNA technologies have great potential as traceability systems for living items because the unique identifier is already present in the DNA sequence. It might be necessary to capture some of the CTE's along with the DNA but each animal in a herd of livestock will have a unique DNA pattern. The methods for DNA sequencing and DNA barcoding are becoming simpler every year and analytical tests are becoming more and more affordable [1]. Many different types of DNA tests exist from very simple to whole genome sequencing which provides complete information. For live animals and meat-based products, DNA technologies offer many opportunities in tracing and this will only increase as technology continues to improve.

Finally, with respect to the distribution and sale of produce, bar codes are being used by the industry with increasing frequency. Bar code stickers can be affixed to a wide-range of produce which contain information relating to the origin and distribution of the product. Here too, these technologies have enhanced the traceability of these products significantly.

Traceability Laws and Regulation

The federal laws and regulations governing the traceability of meat and poultry require companies, as detailed in Chap. 2, to "keep records that will fully and correctly disclose all transactions in their business subject to the [relevant] Act[s]" [7]. Thus, any time any livestock or, more importantly, food products derived from any livestock are sold, the parties must maintain traceability records that describe the article being sold and the nature of the transaction (*i.e.*, bills of sale and lading, invoices, and all receiving and shipping papers). In addition to maintaining sale and distribution records, companies must also maintain records relating to the production, packaging and labeling of any food products they manufacture. The final products must also be labeled with additional information (to include the federal inspection legend, the establishment number of the producing facility, handling instructions and additional information) that is compliant with the relevant federal regulatory standards. See Chap. 2.

In addition to these requirements, the Food Safety and Inspection Service ("FSIS") also requires official establishments and retail stores that grind raw beef products to keep grinding records showing the source of the raw materials being used. See Chap. 2. Under the rules, all establishments and retail stores that grind raw beef products are required to maintain the following records: (1) the establishment numbers of establishments supplying material used to prepare each lot of raw ground beef product; (2) all supplier lot numbers and production dates; (3) the names of the supplied materials, including beef components and any materials carried over from one production lot to the next; (4) the date and time each lot of raw ground beef product is produced; (5) and the date and time when grinding equipment and other related food-contact surfaces are cleaned and sanitized. If FSIS visits a retail facility and finds that it is not in compliance, the agency may tag product or equipment (thus preventing its use), or take action to prevent further non-compliance, such as seizing product or even prohibiting the retail establishment from selling FSIS regulated product altogether.

These requirements give FSIS the necessary tools to trace the distribution of meat and poultry products through the supply chain. Most importantly, these tools allow the agency to more effectively identify, trace, and contain foodborne illness outbreaks when they occur.

The Role of Traceability in Foodborne Illness Investigations

For many years, it was extremely difficult for FSIS to trace foodborne illness outbreaks involving raw animal product back to their original source. Prior to the Jack-In-The-Box Outbreak in 1993, which sickened 600 people and killed four, there was no effective national system in place to effectively identify, track and contain foodborne illness outbreaks as they occurred. The Jack-In-The-Box Outbreak was only identified by healthcare professionals because the victims were located within a relatively limited geographical area. Indeed, many of the victims were being treated in the same hospitals, and some of the victims were even being treated by the same doctors. As a result, the existence of an emerging outbreak was eventually identified, and healthcare providers worked closely with public health officials to determine the ultimate source. The investigation determined that the outbreak was caused by undercooked hamburgers being served by the restaurant chain.

Following the conclusion of that outbreak and investigation, the federal government became concerned that similar outbreaks could be happening throughout the nation without any means to detect them. To enhance the government's ability to detect emerging foodborne illness outbreaks on a national scale, as they were actually occurring, the government implemented a mandatory system of foodborne illness reporting.

Beginning in the late 1990s, this new, mandatory system required that, whenever a doctor in any of the 50 states discovered that a patient had tested positive with a pathogen of concern (such as *Listeria monocytogenes*, *Salmonella*, or *E. coli* O157:H7), he or she was required to report that finding to the relevant state health department. The individual states would then request copies of the isolates obtained from the sick case patients and test them for the specific genetic DNA fingerprint of the pathogen making the patient sick. This is the same mandatory reporting system that remains in place today.

Once a case-patient's genetic DNA fingerprints are obtained by the state, those fingerprints are uploaded to PulseNet, a system designed by the Centers for Disease Control (CDC) to track emerging foodborne illness outbreaks and discussed in Chap. 3. When genetic DNA fingerprints are uploaded by the states from victims, those fingerprints are compared by the CDC. When the fingerprints being uploaded are indistinguishable (*i.e.*, they closely match), the CDC knows that a foodborne outbreak may be emerging. CDC then shares that information with FSIS and other federal, state and local health departments as they work collaboratively to interview the case patients, obtain information about their food exposures, and determine a common source.

When PulseNet first became operational, CDC immediately began identifying numerous outbreaks involving a wide-range of foods. Over the next decade, as the system became more effective and robust, it became clear that raw animal foods, and ground beef in particular, were responsible for a large number of outbreaks. Although PulseNet was extremely useful to FSIS' efforts to pinpoint ground beef products as the cause of many of these outbreaks, the system could not assist FSIS in determining the original source of the contamination that was discovered in those ground beef products.

This is because, in the past, neither retail establishments nor federally-regulated beef processing establishments were required to keep grinding records. Because many retail establishments process or regrind their own ground beef at the retail level, and often use meat products from many different sources or suppliers in each batch, it was impossible in many cases for FSIS to identify which supplier's product introduced the contamination. Additionally, even in those cases where FSIS was able to identify a single processor who's ground beef was contaminated, FSIS could rarely identify the harvest facility from which the contamination originated. This is because virtually all processors, like most retailers, would comingle trim from many different suppliers in each batch of product. Thus, even in those cases where there were records showing the source suppliers used for each batch (in some cases there were not), it was still nearly impossible for FSIS to identify the harvest facility from which the contamination originated.

To enhance its ability to solve outbreaks and trace the offending contamination back to its original source, FSIS has modernized its traceability policy substantially. For starters, as noted above, in addition to the traceability records the regulations have always required companies to maintain, FSIS is now requiring all ground beef processors and retail establishments to maintain grinding records. These records will significantly enhance the agency's ability to trace the original sources of contaminated product.

In addition, FSIS has also enhanced its longstanding investigative philosophy. Under the previous FSIS investigative policy, the agency would in many cases simply record the names of any suppliers whose trim was used to process a positive batch, and then only conduct follow-up inquiries under limited circumstances. Now, when ground beef tests positive for *E. coli* O157:H7 (or any non-O157:H7 STEC), the agency will launch an investigation to find the original source of the contamination. These new investigations are directed at the supplier (or suppliers) of any trim used to process the contaminated batch of ground beef. The agency will direct an Enforcement Investigations and Analysis Officer ("EIAO") to visit each supplier and review the supplier's microbiological testing records. When such a review occurs, if there is any evidence of a lack of system control, wide-spread contamination, or other problems, FSIS may strongly encourage a broad recall of all potentially affected products. This could include, depending upon the circumstances, both trim and intact products that were fabricated during the questionable periods.

It is important to recognize that, from a regulatory standpoint, whether you harvest or process beef, traceability will continue to take on increased importance. As FSIS' focus on food safety continues to mature, the agency will increasingly expect all companies in the meat and poultry supply chain to maintain increasingly robust records showing exactly what they are receiving, exactly what they are processing, and exactly what they are selling.

References

1. Galimberti A, De Mattia F, Losa A, Bruni I, Federici S, Casiraghi M, Martellos S, Labra M (2013) DNA barcoding as a new tool for food traceability. Food Res Int 50:55–63
2. GS1 US (2014) Traceability for meat & poultry U.S. implementation guide. R2.0
3. Liu S, Zhang D, Zhang R, Liu B (2013) Analysis on RFID operation strategies of organic food retailer. Food Control 33(2):461–466
4. McEntire J, Arens S, Bugusu B, Busta F, Cole M et al (2010) Traceability (product tracing) in food systems: an IFT report submitted to the FDA, volume 1: technical aspects and recommendations. Compr Rev Food Sci Food Saf 9(1):92–158. https://doi.org/10.1111/j.1541-4337.2009.00097.x
5. mpXML (2010) Traceability for meat and poultry US implementation guide. Issue 1–14
6. Smith GC, Tatum J, Belk K, Scanga J, Grandin T, Sofos J (2005) Traceability from a US perspective. Meat Sci 71(1):174–193. https://doi.org/10.1016/j.meatsci.2005.04.002
7. Traceability for Livestock Moving Interstate 9 (2013) CFR 320.1. http://www.gpo.gov/fdsys/pkg/FR-2013-01-09/pdf/2012-31114.pdf
8. Zhang J, Bhatt T (2014) A guidance document on the best practices in food traceability. Compr Rev Food Sci Food Saf 13(5):1074–1103. https://doi.org/10.1111/1541-4337.12103

Chapter 7
Fresh Produce

Ed Treacy, Jane Proctor, Dan Vaché, and Angela Fernandez

Abstract Collaborative efforts of leaders in the fresh produce industry in the United States and its major trading partner Canada have played a pioneering role not only in raising awareness of the importance of chain wide traceability by industry and other stakeholders, but also in developing best practices to support industry's implementation. The PTI model and its best practices are being adopted by other fresh food sectors, such as seafood, meat, poultry, dairy, deli and bakery. Created in 2007, the Produce Traceability Initiative (PTI) was the first industry-led, industry-specific, supply chain-wide traceability initiative in the United States. PTI was created to achieve whole chain traceability by incorporating technology with commonly-used product identification standards to create linkages between internal traceability programs. The majority of the companies in the produce industry had very good traceability programs in place within their organizations, but they were not linked, and the relevant traceability information was not transferred or captured as product moved through the supply chain. The adoption of case level labeling based on GS1 Standards with a Global Trade Item Number™ (GTIN®) and Batch/Lot Number (the minimum requirement for the Produce Traceability Initiative) has become the model for several other industries' traceability programs and initiatives. While recent regulatory developments have impacted PTI implementation, companies are finding that PTI compliance provides substantial business benefits and savings beyond traceability.

Ed Treacy (✉)
Produce Marketing Association, Newark, DE, USA
e-mail: etreacy@pma.com

J. Proctor
Canadian Produce Marketing Association, Ottawa, ON, Canada
e-mail: jproctor@cma.ca

D. Vaché
United Fresh Produce Association, Redmond, WA, USA
e-mail: danvache@comcast.net

A. Fernandez
GS1 US, Lawrenceville, NJ, USA
e-mail: afernandez@gs1us.com

J. McEntire, A. W. Kennedy (eds.), *Food Traceability*, Food Microbiology and Food Safety, https://doi.org/10.1007/978-3-030-10902-8_7

Keywords PTI · Produce traceability initiative · Standards · Whole chain traceability

The Genesis

The fresh produce industry in the United States was abruptly awakened by a massive foodborne illness outbreak linked to spinach in 2006. One single spinach packaging operation in California unknowingly received, processed and shipped packages of fresh spinach in August 2006 from a ranch growing organic spinach that was contaminated with *E. coli* O157:H7.

When the U.S. Food and Drug Administration (FDA) warned consumers on September 14, 2006 to "not eat bagged fresh spinach," sales and shipments of all fresh spinach virtually stopped [1].

The contaminated spinach shipped into the supply chain caused illness in over 200 individuals and took the lives of four people: Ruby Trautz, age 81, from Nebraska; Marion Graff, 77, from Wisconsin; June Dunning, 86, from Maryland; and Kyle Allgood, 2, from Idaho.

In addition to the tragic loss of human life, consumers in the United States suddenly lost confidence in the produce industry's ability to trace its products. American consumers had previously assumed the produce industry was capable of locating any product, any time in the supply chain, and had the ability to trace shipments and swiftly recall product. With the spinach outbreak lasting over 2 months, consumers lost that confidence.

This led to the third major impact of the spinach outbreak: the financial impact. Industry losses in 2006 were estimated to be more than $74 million. In 2007, the losses grew to $350 million, and sales recovered to only 80% of the pre-recall levels. It took another 4 years for the spinach sector to recover to 2006 sales volume. Many small- and medium-size businesses were not able to survive these losses, and businesses that had survived through multiple generations of family ownership went out of business – even though their spinach had not caused this outbreak.

A New Traceability Vision: Interoperability

This outbreak and the four deaths it caused awakened the produce industry. Leaders knew that past industry practices had to change to better safeguard public health. They also knew that the U.S. federal government was contemplating an update to the Food Drug and Cosmetic Act. While the 2002 Bioterrorism Act required that food companies be able to trace their products one step forward and one step back, the produce industry started to question if this was now insufficient for safeguarding consumers and businesses alike. While most produce companies already had very robust internal traceability systems within their companies at that time, their traceability programs were not interoperable with their trading partners. Each trading partner identified product with its own internal identifiers or item numbers. As the majority of produce is sold unpackaged, there was no Universal Product Code (UPC) or other common identifier on the product itself to assist regulators in their traceback investigation. These challenges lengthened the time it took regulators to conduct traceback investigations, as they had to validate the information each trading partner provided to ensure they were tracing the same product.

Two major produce industry initiatives were created as a direct result of the 2006 spinach outbreak: the California Leafy Green Products Handler Marketing Agreement and the Produce Traceability Initiative.

The California Leafy Green Products Handler Marketing Agreement (LGMA) was established in the spring of 2007. The LGMA, operating with oversight from the California Department of Food and Agriculture, provides a mechanism for verifying that farmers follow established food safety practices for lettuce, spinach, and other leafy greens. This model was replicated in Arizona, the other major leafy green growing region [3].

The Produce Traceability Initiative (PTI) was created in 2007 as a voluntary industry-led initiative to develop standards and protocols to facilitate traceability in the U.S. fresh produce supply chain. The boards of directors for Produce Marketing Association (PMA), Canadian Produce Marketing Association (CPMA), and United Fresh Produce Association (United Fresh) recognized the need for greater progress in implementing a consistent, whole chain traceability solution, and made supporting implementation of traceability a priority of each association [10].

The three associations worked together to recruit members to an initial PTI Steering Committee. The first meeting of the committee was held on January 9, 2008 in Atlanta, Georgia.

Cathy Burns, then chief operating officer of the grocery retailer Food Lion, was elected chair of the PTI Steering Committee. The original PTI Steering Committee included:

- Six foodservice companies: Amerifresh, Markon Cooperative, Inc., McDonald's, Pro*Act, Sysco Corporation and U.S. Foodservice;
- 11 grocery retailers: Food Lion, H-E-B, W. Newell & Co. (Supervalu), The Kroger Company, Loblaw, Safeway Stores, Inc., Publix Super Markets, Schnuck Markets, Inc., Sobeys, Inc., Wal-Mart Stores, Inc., Wegmans Food Markets; and
- 20 produce suppliers: AEPQ (Quebec Apple Packers), B.C. Tree Fruits Limited, Ballantine Produce Co., Inc., C.H. Robinson Worldwide, Inc., Dole Food Company, Inc., Domex Superfresh Growers, Driscoll's, Duda Farm Fresh Foods, Inc., Fresh Express, Inc., Fresh Innovations, LLC., Frontera Produce, Ltd., L&M Companies, Mann Packing, Naturipe Farms, LLC., Procacci Brothers, The Oppenheimer Group, Pandol Brothers, Inc., Ready Pac Produce, River Ranch Fresh Foods, LLC., and Tanimura & Antle.

The PTI Action Plan, Founded on GS1 Standards

The PTI Steering Committee created an Action Plan to help the produce industry to adopt an effective, whole-chain traceability program. That Action Plan incorporated the use of technology with the use of common and globally unique product identification standards to create linkages between companies' internal traceability programs from the point the fresh produce is packed for shipping until it is delivered to a retail store or foodservice establishment. The Steering Committee reached agreement on four key elements for implementing industrywide traceability best practices.

First, the group confirmed that use of GS1 Standards, developed by the international standards organization GS1 – the most efficient worldwide approach to achieve system-wide (i.e., both internal and external) traceability – should be adopted by the produce industry. GS1's traceability standards were based on standards developed in 2004 by the Canadian Initiative Can-Trace, a collaborative and open initiative committed to developing traceability standards for all food products sold in Canada. The Can-Trace standards were then foundational in the development of the GS1 Global Traceability Standard and the Global Fruit & Vegetable Traceability Implementation Guide [11].

Second, it was agreed that creating an industry timeline for adoption of standards was needed. Steering Committee participants agreed to begin evaluating what might be required to implement GS1 Standards within their own operations and reported back at the next Steering Committee meeting their recommended implementation timelines.

Third, the committee agreed to discuss ways in which companies could best show their support and commitment to adoption.

Fourth, the Steering Committee agreed that traceability standards should be adopted at the case level initially, as the backbone of supply chain traceability, since virtually all of fresh produce is shipped in cases. Case level, in this definition, includes cases, cartons, boxes, flats, returnable plastic containers (RPCs) and bins.

Another Outbreak Accelerates the Initiative

In the summer of 2008, another foodborne illness outbreak occurred. The investigation linked the cause to fresh produce. Originally thought to be caused by tomatoes, the culprit was later identified to be fresh jalapeño and serrano peppers from Mexico.

In July of 2008, Bryan Silbermann, then PMA CEO, testified before the House Agriculture Committee's Horticulture and Organic Agriculture Subcommittee, as part of a hearing that was called to study produce traceability because of claims that problems had slowed the still-lingering *Salmonella* Saintpaul foodborne illness investigation linked to fresh produce. "The produce industry has already rapidly changed to avoid the introduction of risk into the food system," because of its longtime commitment to food safety and the recent impetus provided by the food-borne illness outbreak linked to spinach in late 2006, Silbermann told committee members. "It is not the private sector's role to wait passively for government to regulate; we must act, and we are doing so" [7].

This outbreak lasted 3 months. The Centers for Disease Control and Prevention (CDC) reported that more than 1442 persons were infected with *Salmonella* Saintpaul of the same genetic fingerprint in 43 states, the District of Columbia, and Canada. CDC also reported that at least 286 persons were hospitalized, and the infection might have contributed to two deaths [5].

At the third meeting of the PTI Steering Committee, seven milestones were created to assist the industry with implementing PTI:

1. Brand owners must obtain a GS1-issued Company Prefix.
2. Brand owners must assign 14-digit Global Trade Item Numbers (GTINs) to all case configurations. The Steering Committee highly recommends that companies use the number assignment strategy already created by the trade associations to minimize the number of GTINs created and to allow for consistency across industry segments.
3. Brand owners must provide and maintain their GTIN information (and corresponding data) to their buyers.
4. All parties must have the systems to capture and store GTINs and subsequent information.
5. Those parties packing the product are responsible for providing the GTIN, lot number and pack/harvest date in a human-readable form on each case (Note: Pack/harvest date is optional if it is already embedded in the lot number).
6. Those parties packing the product are responsible for encoding the GTIN, the batch/lot number, and the pack/harvest date in a GS1–128 barcode. (Note as above: Pack/harvest date is optional if already embedded in the lot number).
7. Each handler of the case must read and store the following information both one step up and one step down the supply chain: GTIN, batch/lot number, pack/harvest date (if not already included in the batch/lot number), shipper ID; shipper name, shipper address, receiver ID, receiver name, receiver address, date of shipment, date of receipt, quantity, unit of measure, and shipment ID.

PTI Officially Launches

The PTI Action Plan was officially launched to industry in October 2008. Thirty-four companies from across the produce supply chain immediately endorsed the new plan to move the industry to embrace a common approach for electronic produce traceability by the end of 2012.

The plan aimed to maximize the effectiveness of the industry's current traceability procedures, to improve internal efficiencies, and to assist public officials when they need to quickly trace back a product. Intended to enhance overall supply chain traceability in speed and efficiency, the standardized system would significantly improve the industry's ability to narrow the impact of potential recalls, product withdrawals, or other events necessitating the quick identification of product on its journey from field to the consumer.

The Action Plan involved adopting a standardized system of case bar coding at the batch or lot number level, for all produce sold in the United States to allow product to be tracked throughout the distribution chain. The lot/batch level was chosen because virtually all packers of produce were already using a batch or bot number in their own recordkeeping.

Target dates were assigned to each milestone:

1. 2009 – Obtain a Company Prefix from GS1
2. 2009 – Assign GTINs to Cases
3. 2009 – Provide GTIN Information to Buyers
4. 2011 – Show Human-Readable Information on Cases
5. 2011 – Encode Information in a Barcode
6. 2011 – Read and Store Information on Inbound Cases
7. 2012 – Read and Store Information on Outbound Cases

In February of 2010, the PTI Steering Committee – now more than 50 members strong – met in Dallas. During that meeting, GS1 US was officially added as a fourth supporting association, in addition to PMA, CPMA and United Fresh.

At that meeting, a new governance structure was adopted based on the governance model of the Foodservice GS1 US Standards Initiative that was already gaining momentum at the time. The Steering Committee was replaced with a Leadership Council, an Executive Committee, four original working groups and a Commodity Interest Group. A Buyers Working Group was added in 2013 (See Fig. 7.1)

The newly formed PTI Leadership Council was chaired by Cathy Burns, who was now president of the Food Lion family of companies.

The PTI Leadership Council included:

- seven retailers: Delhaize America/Food Lion, The Kroger Company, Publix Super Markets, Safeway, Supervalu, Wakefern Food Corporation and Wal-Mart Stores;
- 15 suppliers represented were; C.H. Robinson Worldwide, Chiquita/Fresh Express, Del Campo Supreme, Del Monte, Dole Food Company, Driscoll Strawberry Associates, Frontera Produce, Jem-D International, L&M Companies, Paramount Citrus, Sunkist Growers, Tanimura & Antle, Taylor Farms, The Oppenheimer Group and Wada Farms Marketing Group;
- four wholesaler/terminal market operators: Castellini Company, D'Arrigo Brothers Company of NY, Four Seasons Produce and Liberty Fruit Company; and
- five foodservice operators: Darden Restaurants, Markon Cooperative, PRO*ACT, Sysco Corporation and US Foods; and

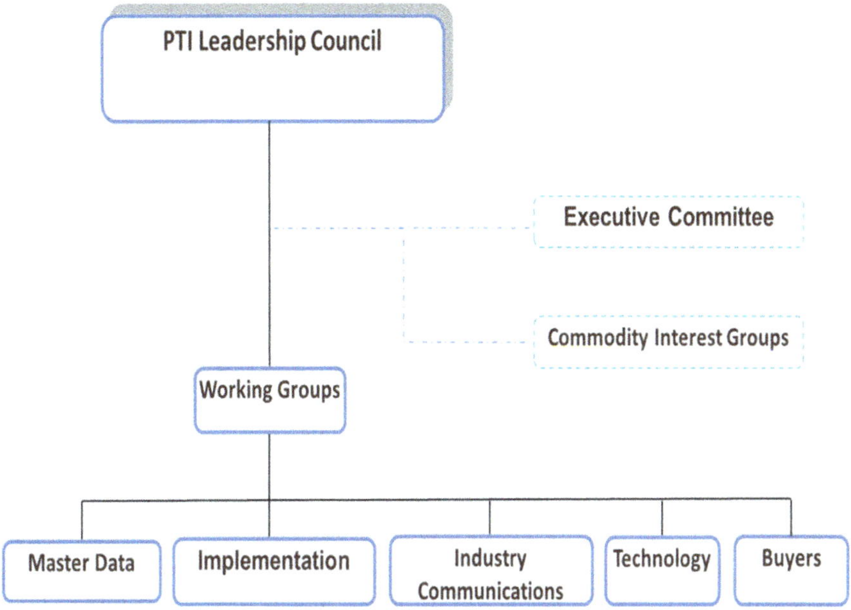

Fig. 7.1 PTI governance structure

- five industry associations: the four PTI administering associations PMA, CPMA, United Fresh, and GS1 US, joined by Food Marketing Institute (FMI).

With a new governance structure now in place, objectives were established for each of the working groups to help encourage industry adoption and implementation of PTI best practices:

- Implementation Working Group: Guide and promote industrywide adoption of GS1 Standards as the foundation of the PTI, including by developing best practices, identifying solutions to implementation issues, and tracking industry implementation.
- Master Data Working Group: Address issues regarding product attributes identification and communication of product data between trading partners.
- Technology Working Group: Provide a forum for technology providers to collaborate to support the initiative.
- Industry Communications Working Group: Ensure two-way communication between the initiative and industry.

In addition to the four working groups, a Commodity Interest Group was formed to provide a forum for regional associations and commodity groups to engage and support the initiative.

To participate in a working group, an industry representative must be a member of at least one of PTI's four administering organizations: PMA, CPMA, United

Fresh or GS1 US. Participation for representatives of technology company, service providers or consultants was restricted to the Technology Working Group.

A fifth working group, the Buyers Working Group, was created in January 2013 to develop best practices and provide technical and functional expertise from the buyer's perspective.

Each working group is co-chaired by a representative from the produce industry and a staff member of one of the four administering associations:

- Implementation Working Group Co-chairs: Steve Roosdahl of The Oppenheimer Group and Ed Treacy of PMA
- Master Data Working Group Co-chairs: Tom Casas of Tanimura & Antle and Scott Brown of GS1 US
- Technology Working Group Co-chairs: Todd Baggett of Redline Solutions and Andy Kennedy of FoodLogiQ, who was replaced by Julie McGill of FoodLogiQ when Kennedy became the acting director of the Global Food Traceability Center and Dan Vaché of United Fresh, who was replaced by Jane Proctor of CPMA, when Vaché left United Fresh.
- Communications Working Group Co-chairs: Sabrina Pokomondy of Jem-D International (now known as Red Sun Farms) and Julia Stewart then of PMA co-chaired the Industry Communication Working Group. When Stewart left PMA, she was replaced by Krisztina Vida of GS1 US.
- Commodity Interest Group Co-chairs: Joel Nelsen of the California Citrus Mutual and Jane Proctor of CPMA
- Buyer Working Group Co-chairs: Teri Miller of Food Lion/Delhaize America and Ed Treacy of PMA.

The working groups engaged immediately, meeting regularly to create best practice and guidance documents to be posted to the PTI website at http://www.produce-traceability.org/.

As of this writing, the following documents have been created to assist industry companies to implement PTI [8].

- Best Practice for Use of Hybrid Pallet Labels by Receivers
- Best Practices for Private Label/Brand
- Best Practices for Repacking/Commingling
- Best Practice for Produce Brokers
- Best Practices for Direct Print
- Best Practices for Formatting Case Labels
- Best Practices for Preparing to Assign GTINs
- Best Practices for Preparing to Assign GTINs (Spanish)
- Best Practices for Product Substitutions
- Best Practices for Cross Docking
- Best Practices for Cross Docking (Spanish)
- Best Practices for Formatting Hybrid Pallets Labels
- Guidance on Benefits of Advance Ship Notice versus Hybrid Pallet Labels

- Why and How to Use EDI 856 Advance Ship Notice/Manifest Transaction
- Guidance on Choosing a Technology Provider
- Best Practices Data Synchronization
- Data Synchronization: Grade Codes
- Data Synchronization: Commodity and Variant Codes
- Data Synchronization: Growing Method Codes
- Data Synchronization: Units of Measure Codes
- Data Synchronization: Package Type Codes
- Data Synchronization: Country of Origin Codes
- Data Synchronization: Worksheet Example
- Data Synchronization Template
- PTI Checklist for Receivers/Buyers
- PTI Checklist for Growers/Packers/Shippers
- Voice Pick Code Calculator (Source: HarvestMark)
- Guidance for Sharing Traceback Data***
- Guidance for GLN Assignment

**** FDA staff participating in the working group that developed this Guidance document*

Moving from Action Plan to Implementation

Several working group members engaged in pilot projects to validate and create PTI best practices. PTI adopted the Global Trade Item Number (GTIN) and the Batch/Lot Number as the two pieces of common information needed for whole-chain traceability. PTI requires that this information be printed on a case label in two forms: a machine-readable GS1-128 barcode, and human-readable text. The pack date or expiry date was agreed to be considered optional, after much debate amongst the working groups.

A critical barrier for distributors to implement PTI was identified soon after the working groups had been created and deployed: how to capture the information on the barcoded label at the time of case selection/order picking, for companies that utilize voice-directed picking systems.

Voice-directed picking systems do not use case-code scanners, as all of the commands are translated to voice prompts conveyed through headsets worn by order selectors. If order selectors were required to scan the barcode on the PTI label, their productivity would have decreased dramatically. This decrease in productivity would require large investments in additional labor and building expansions across retail and foodservice distribution centers.

Fortunately, Dr. Elliott Grant, founder and chief technology officer of HarvestMark Technologies and a member of the PTI Technology Working Group, came up with the idea of using the CRC-16 hash computation algorithm to translate the GTIN and Lot Number into a 4- digit number. That 4-digit number became known as the PTI Voice Pick Code. When prompted, the PTI Voice Pick Code can be read into a voice-directed picking system, so that the correct GTIN and Lot Number can be electronically assigned to the case being picked. This concept was successfully tested, and then adopted as a required standard for PTI. Dr. Grant obtained a patent on the business process of using the algorithm for this purpose and turned over the patent to the industry to use at no cost, ensuring that no one could profit from this business solution.

To Trace It, Put a Label on It

With the required minimum data elements now agreed to by PTI's working groups, the groups then set about developing PTI-compliant standardized labels. The first label created was a case label (Fig. 7.2) for use on all cartons, boxes and containers except RPCs. Three pieces of information were deemed for this case label:

- The machine-readable GS1-128 barcode which incorporated the GTIN and Batch/Lot Number;
- The human-readable GTIN and Batch/Lot Number; and
- The four-digit PTI voice pick code.

The second standardized label created was the Hybrid Pallet Label (HPL) as shown in Fig. 7.2. This label was created for companies not able to utilize the EDI transaction Advance Ship Notice (ASN) (further described in Chap. 11). The HPL was designed to eliminate the need for receivers to scan every case on receipt. It simply provided a summary of the barcodes on the cases on the pallet (Fig. 7.3).

The third label created was for use on RPCs (Fig. 7.4). The RPC label required additional information as all of the required information for trade and regulatory purposes had to be incorporated into one label. With corrugated cases, information such as country of origin and other regulatory information could be printed on the case. The introduction of the RPC PTI label was welcomed by the industry, as it replaced the expensive 2 ½-inch × 11-inch wrap around label required by three of the four largest retailers using RPCs.

Fig. 7.2 Example of a PTI compliant label

Fig. 7.3 Example of a hybrid pallet label

Fig. 7.4 Example of a PTI complaint label for use on reusable plastic containers

FSMA Formalizes Recordkeeping Requirement

The U.S. Food Safety Modernization Act (FSMA) was signed into law on January 4, 2011. In addition to mandating a regulation be developed for addressing record-keeping for high risk foods, FSMA also required FDA to conduct two traceability pilot projects, one with fresh produce and one with processed food. FDA was required to report back to Congress their recommendations based on the two pilot projects' results by July 4, 2011. FDA contracted with the Institute of Food Technologists (IFT) to perform the pilots [4]. PMA, GS1 US and United Fresh, along with many produce industry members, volunteered to work with FDA to accomplish the produce pilot.

Twelve different traceback scenarios were created. One of these scenarios was a fully PTI compliant scenario; that scenario completed a full traceback and determined convergence in less than 24 h. This validated that the goal of PTI was achievable: to be able to complete end-to-end trace-back and trace-forward investigations within 24–48 h [4].

The leaders of the Produce Traceability Initiative have worked closely with FDA to keep them informed of its industry-led efforts and have encouraged the agency to develop traceability regulations that are supportive of PTI. Michael Taylor, then the FDA Deputy Commissioner for Foods, acknowledged the efforts of PTI during a May 5, 2011 PTI Leadership Council Meeting: "We are keenly aware that industry has been at the forefront of understanding traceability, and in order to make progress we know we are going to need to build on and embrace the work that industry has done" [2].

The produce industry continued to be struck by outbreaks and recalls after the 2006 spinach recall. In addition to the 2008 jalapeño and serrano peppers event, a third major foodborne outbreak occurred in 2011. This was caused by *Listeria monocytogenes* on cantaloupes from a farm in Colorado [6] this outbreak caused 33 deaths and one miscarriage. In addition to being the worst death toll related to a foodborne illness in the United States in many decades, this outbreak resulted in the grower/shipper, its marketing company and its retailer customer all being sued for introducing into commerce cantaloupes that were the cause of the deaths. Many in the industry believe the severity of the impact of this outbreak would have been reduced if the entire produce supply chain were PTI compliant. The inability of the FDA and CDC to conduct a rapid traceback investigation, identify the source, and subsequently trace forward to determine where the affected product had been distributed, essentially shut down the entire U.S. cantaloupe industry. While having traceability in place doesn't reduce the risk of foodborne illness, it can play an important role in determining the scope of implicated product and subsequently removing product from the supply chain once a food safety event is identified. Efficient and expedient whole-chain traceability will also assist regulators in quickly identifying the products that are not involved with a food safety event.

Major Retailers Give PTI a Boost

The first retailer to call for PTI compliance was Publix Super Markets. Publix requested that its suppliers be PTI compliant starting in the spring of 2012. Publix had redesigned their receiving and quality inspection processes to utilize the bar-coded PTI label in order to integrate their receiving and quality inspection processes. In so doing, Publix also reported that they realized major labor savings and reduced processing times in their produce distribution centers.

The next major retailers to request PTI labeling from their suppliers was Walmart and Sam's Club. On May 29, 2013, they issued a letter to their suppliers requesting compliance by November 1, 2013. They also requested that the RPC PTI label be used on all cases and RPCs. This request from the country's largest grocery retailer was what many in the industry were waiting for; it meant that PTI was here to stay and was on its way to becoming the industry standard.

Walmart's action also caused many conversations amongst executives at most of the other grocery retailers. They wanted to know what benefits Walmart and Sam's Club were going to accrue from their implementation, and was it going to put their company at a disadvantage. Walmart was very open about their expectations. For example, at the time they were in the practice of writing the received date on a sticker on each case of produce received into their stores. With the pack date now imprinted on the PTI label, Walmart and Sam's Club would save the labor and the cost of having to affix their own sticker – and more importantly, they could now rotate product on the sales floor based on the pack date, not the date the product was received into the store as had been their previous practice. Executives from Walmart reported that this change was having a positive measurable impact on their quality.

In June 2014, at the United Fresh conference in Chicago, Mike Agostini then of Walmart revealed that the second impetus for its implementation of PTI and its standardized label came from Walmart's legal department after the company was named in a lawsuit stemming from the 2011 cantaloupe foodborne illness.

The third major retailer to require PTI case labeling was Whole Foods. They informed their suppliers to label all cases bound for Whole Foods with the RPC PTI label. This label was consistent with Walmart's label requirements. Whole Foods was the first retailer to require suppliers to provide the information on the PTI RPC labels – that is, the GTIN and Lot Number – electronically in the form of an Advance Ship Notice (ASN). The use of ASNs had been pilot tested by member companies of the PTI Implementation Working Group and had proven to be the preferred method for communicating the GTIN and Batch/Lot Number.

Early Adopters See Business Benefits

These three major retailers' requirements for PTI labeled cases solidified that PTI was being adopted as the industry standard. This was welcomed by the early adopters of PTI in the supplier community. Some suppliers had begun labeling their cases before they were requested to do so by their customers.

Beyond speeding traceability, some of these innovative early adopters realized other benefits from applying a label with a machine-scannable barcode. Some companies used this label in their claims process to determine which packer or picking crew was responsible for the product for which a claim was being made and used this information to improve their processes. Some companies integrated scanning of the PTI label into their production tracking processes to replace the manual tracking of piecework. One company paid for its entire PTI implementation by printing the country of origin on the PTI label, reducing the need to keep separate inventories of cardboard boxes with "Product of USA" and "Product of Mexico" printed on the box. One of the most creative versions of the label was a Mexican company that incorporated a picture of their employee who packed the case on the PTI case label. They reported that their quality increased and claims decreased, and that had paid for the company's entire implementation cost.

Foodservice, Blockchain, Blockchain Projects Also Shine Light on PTI

As of August 2018, estimates of the percentage of produce cases that are labeled with PTI compliant labels are over 60%. The companies currently applying PTI compliant labels include large, medium and small packers, re-packers and re-graders across all produce commodities.

The concurrent Foodservice GS1 US Standards Initiative and GS1 US Retail Grocery Initiative also helped spur implementation of PTI within the produce industry. Many foodservice operators began requiring suppliers to label all cases with a GS1-128 barcode to facilitate scanning cases at their restaurants' store level. The foodservice operators requiring the GS1-128 barcode on their cases 1 as of August 2018 are: ARCOP (Arby's), Chick-fil-A, Chipotle Mexican Grill, CKE Restaurants, Dine Brands Global (IHOP, Applebee's), Domino's Pizza, IPC (Subway), Panda Restaurant Group, QSCC/Wendy's and YUM! Brands.

Enter blockchain. A collaborative pilot project evaluating blockchain's value that includes Dole, Driscoll's, The Kroger Company, Walmart, Wegmans, IBM and other companies, was announced in mid-2017. These pilot tests initially focused on traceability and produced very encouraging results, proving that a full traceback can be accomplished in seconds rather than days if all partners in the supply chain share their data. PTI case labeling is a foundational requirement of these pilots, which have raised awareness about and of the value of the Produce Traceability Initiative.

FDA Sees Value in PTI

The deputy commissioner for foods and veterinary medicine for the Food and Drug Administration, Stephen Ostroff, said PTI labels make it easier for the agency to reconstruct supply chains in the course of investigating outbreaks in a May 31, 2018 article published in the *The Packer* newspaper. "The short answer is yes," Ostroff said. "The ability for us to do a traceback and the ability for us to do a traceback quickly is directly dependent on what information is available and the ease of access of that information, and so (PTI) I think helps to address that" [9].

What's Next for PTI?

The produce industry continues to anxiously await FDA's release of implementing regulations to fulfill FSMA requirements on record keeping. The PTI community believes that implementation will increase substantially when these regulations are finalized, and interest in traceability will spur further adoption of PTI and its Best Practices, ultimately benefitting the industry and consumers not only in the United States, but around the world.

References

1. FDA warning on serious foodborne *E. coli* O157:H7 outbreak one death and multiple hospitalizations in several states [Internet]. Silver Spring: US Food and Drug Administration; c2006 [cited 2015 Nov 3]. Available from: http://www.fda.gov/NewsEvents/Newsroom/PressAnnouncements/2006/ucm108731.htm
2. FDA participation, pilot projects, benchmarking report are highlights of May 2 PTI Leadership Council meeting [Internet]. [Location unknown]: Produce Traceability Initiative; c2011 [cited 2015 Nov 4]. Available from: http://www.producetraceability.org/press-contact/details/fda-participation-pilot-projects-benchmarking-report-are-highlights-of-may
3. LGMA [Internet]. Sacramento: California leafy green products handler marketing agreement; c2015 [cited 2015 Nov 3]. Available from: http://www.lgma.ca.gov/
4. McEntire J, Bhatt T (2012) Pilot projects for improving product tracing along the food supply system [Internet]. Chicago: Institute of Food Technologists. [cited 2015 Apr 27]. Available from: http://www.fdagov/downloads/Food/GuidanceRegulation/UCM341810pdf. Accessed 13 Apr 2015
5. Multistate outbreak of salmonella saintpaul infections linked to raw produce (FINAL UPDATE) [Internet]. Atlanta: CDC; c2008 [cited 2015 Nov 4]. Available from: http://www.cdc.gov/salmonella/2008/raw-produce-8-28-2008.html
6. Multistate Outbreak of listeriosis linked to whole cantaloupes from Jensen Farms, Colorado [Internet]. Atlanta: CDC; c2012 [updated 2015 Sep 4, cited 2015 Nov 4]. Available from: http://www.cdc.gov/listeria/outbreaks/cantaloupes-jensen-farms/index.html
7. Produce industry already enhancing traceability on its own, PMA's Silbermann tells congress [Internet]. [Location unknown]: Produce traceability initiative; c2008 [cited 2015 Nov 4]. Available from: http://www.producetraceability.org/press-contact/details/produce-industry-already-enhancing-traceability-on-its-own-pmas-silbermann
8. Resources and tools [Internet]. [Location unknown]: Produce traceability initiative; c2015 [cited 2015 Nov 4]. Available from: http://www.producetraceability.org/resources/#3
9. The Packer (2018) Is PTI worth the work? FDA, Walmart among those saying yes. [online] Available at: https://www.thepacker.com/article/pti-worth-work-fda-walmart-among-those-saying-yes. Accessed 31 Aug 2018

10. The produce traceability initiative [Internet]. [Location unknown]: Produce Traceability Initiative; c2015 [cited 2015 Nov 3]. Available from: http://www.producetraceability.org/
11. Traceability for fresh fruits and vegetables implementation guide [Internet]. Lawrenceville: GS1 US; c2010 [2015 Nov 4]. Available from: http://www.producetraceability.org/documents/Global_Traceability_Implementation_Fresh_Fruit_Veg.pdf

Chapter 8
Seafood

Barbara Blakistone and Steven Mavity

Abstract Traceability in seafood is reviewed as a commodity under multi-agency regulation and therefore those laws, rules, and policies offer a check on chain of custody. The Bioterrorism Act of 2002 was the first formal act of Congress to require records of previous and subsequent source of food. The seafood industry, mindful of coming traceability requirements in the Food Safety Modernization Act of 2011, prepared its own guide to making a traceability plan and advocates GS1 standards as the preferred tool for track and trace in the supply chain. Further work verified the functionality of the guide, and at this writing, the guide is being applied to unique challenges in the industry such as sustainability, illegal, unreported, and unregulated (IUU) fishing, and seafood fraud. Seafood must join with the other commodities in creating interoperable communication of product tracing information for accuracy, efficiency, and consistency. Seafood is poised to lead the way in traceability because it is a globally traded commodity sold shelf stable, frozen, and fresh.

Keywords Block chain · FDA · Food safety · GS1 · Illegal fishing · NOAA · Fraud · Recall · Seafood · Interoperability · Sustainability · Traceability

Government Compliance

Seafood is a well-regulated industry with a number of agencies (principally the Food and Drug Administration (FDA), U.S. Department of Agriculture (USDA)/ Agricultural Marketing Service (AMS), Customs and Border Protection (CBP), and Fish & Wildlife Services (FWS)) requiring compliance. While not termed traceability, many of the agencies' laws, regulations, and policies offer a check on chain of

B. Blakistone
Packaging Matters, Bothell, WA, USA
e-mail: bblakistone1@aol.com

S. Mavity (✉)
Bumble Bee Seafood, San Diego, CA, USA
e-mail: Steven.mavity@bumblebee.com

J. McEntire, A. W. Kennedy (eds.), *Food Traceability*, Food Microbiology and Food Safety, https://doi.org/10.1007/978-3-030-10902-8_8

custody. As discussed in Chap. 2, several agencies have specific traceability regulations. Below is an overview of several important agencies in the regulation of seafood, highlighting the relationship between these regulations and traceability, whether explicit or implicit.

FDA is the primary agency that regulates seafood. In addition to the application of the Bioterrorism Act-related recordkeeping requirements discussed in Chap. 2, seafood is more specifically regulated by FDA compared to other foods. In December 1997 seafood became regulated by FDA through the Procedures for the Safe and Sanitary Processing and Importing of Fish and Fishery Products. The Guidance document issued by the agency is usually referred to as Seafood HACCP (Hazard Analysis and Critical Control Points) [16], and it became the definitive reference for the industry and much later, during the Food Safety Modernization Act (FSMA) [6] implementation, a model for the food industry. Preparing a traceability plan is much like preparing a HACCP plan. While HACCP is concerned exclusively with food safety in the manufacturing process, traceability has multiple applications in seafood. Traceability marks transitional events (Critical Tracking Events or CTEs) (i.e., transformation, transportation, and depletion) not only in the internal manufacturing process but also with incoming and outgoing materials and products throughout the supply chain. HACCP is a good model for outlining the traceability network.

USDA/AMS requires retailers to mark seafood with country of origin labeling (COOL) as well as a designation of the method of production (i.e., wild capture or farmed) by label, sign, or placard. COOL regulations [11] apply only to raw and unprocessed species, fresh or frozen, so processed seafood items in the supply chain are not covered by these regulations. Any supplier that provides a covered seafood product for sale in a retail operation must maintain records of the immediate previous source and immediate subsequent recipient of the product. The records must substantiate claims and be maintained for 1 year from date of transaction. The initial supplier of the product must keep the records that support the country of origin and method of production claims. Geographical origin and method of harvest are data elements that have application beyond traceability for food safety as discussed later.

CBP exacts duties on goods from specific exporting countries and requires filings that may be helpful in following the supply chain. This agency offers a voluntary program called C-TPAT (Customs-Trade Partnership against Terrorism). The intent is to secure the supply chain. Over 10,000 companies, including seafood companies, belong to C-TPAT, and these companies supply more than 50% of all U.S. imports in value. As the name states, the program is about counter-terrorism, but there are many commonalities in securing the supply chain against terrorism and in traceability. CBP also has COOL rules that require any article, including seafood, of foreign origin entering the U.S. to be marked with the country of origin in order to inform the ultimate purchaser.

Fish and Wildlife Services (FWS) is an agency concerned primarily with preventing the poaching of endangered species. However, FWS requires permits on species like squid, cuttlefish, and octopus, and the permitting process includes naming the country of origin.

In December 2016 NOAA published a final rule known as the Seafood Import Monitoring Program (SIMP) [15] in response to the U.S. Government Task Force on Combating Illegal, Unreported, and Unregulated Fishing and Seafood Fraud report. The rule established traceability monitoring and record keeping requirements in an effort to provide protection against illegal, unreported and unregulated seafood harvest to protect consumer safety, national security, human rights and fair trade. Seafood supply chain companies must provide key data from the point of harvest to the point of importation into the United States.

Acts of the Congress and Traceability

Bioterrorism Act of 2002 [17]

The key elements of the recordkeeping regulation that stemmed from the Bioterrorism Act of 2002 were discussed in Chap. 2. Of note to the seafood industry is the exemption of non-processing fishing vessels, though these vessels must make records available to FDA upon request.

Food Safety Modernization Act (FSMA) of 2011

Section 204 of FSMA on Enhancing Tracking and Tracing of Food and Recordkeeping requires FDA to develop an enhanced traceability system, compatible with domestic and international commerce, which is based on the results of one or more pilot projects that have now been conducted using foods that were subject to outbreaks in the last 5 years. The pilot exercises were required to monitor effective traceability of designated foods while evaluating costs and benefits as well as feasibility of technological tools for tracing. FDA contracted with the non-profit Institute of Food Technologists to conduct the pilot studies. Seafood was not an ingredient or product tested. However, representatives of the seafood industry participated in the deliberations related to the study and ensuing report and recommendations [12].

In the 2013 final report of the pilot, IFT recommended that FDA establish a uniform set of recordkeeping requirements for all FDA-regulated foods that include Critical Tracking Events (CTEs) and Key Data Elements (specific event data or KDEs) as determined by FDA [12]. The results of the pilot will also guide FDA's new recordkeeping requirement for enhanced tracking and tracing of food products by identifying key data elements that are needed to trace a product back through the distribution system. The seafood industry is mindful of the approach and terminology used in these pilots in work it does on traceability.

FSMA stipulated that, in keeping with the recordkeeping requirements of the Bioterrorism Act of 2002, records of the immediate recipient are required to be available within 24 h. Fishing vessels, other than processing vessels, will be exempt. Those that process must keep records of previous source and immediate recipient. FSMA specifically does not require full pedigree (i.e., a sequential addition of "track-forward" records from water to table). Records are not required to the case level. While noting the need for recordkeeping on high risk foods, the FSMA notes that such foods may be subtracted from the list depending on the current safety risks of the food. It seems likely that specific foods will not be named as FDA is more focused on categories that influence risk. At least one seafood species has had recalls for foodborne illness [3].

As of this writing Proposed Rules on Traceability have not been released by FDA, but terms such as CTEs and KDEs mentioned above are now common terms in the food industry, so seafood anticipates that these will be terms in rulemaking by the agency. In anticipation of the Trace Rule, the seafood industry has prepared a guideline as discussed in the Industry Initiatives section below [13]. Due to factors other than food safety, seafood may be subject to more extensive rules than those commodities impacted by FSMA and regulated solely by FDA. See further discussion below under "IUU Fishing."

Applications of Traceability

Traceability in the seafood industry has been accomplished through a variety of methods which vary in complexity and cost and have been generally implemented to meet a particular business need for the user. In every case, identification of discreet quantities of seafood or raw materials as "lots" or "batches" and an understanding of where those lots or batches might change through transformation events, such as co-mingling, processing or combination, serve as the basis for developing the needed records for demonstrating traceability.

For a simple low volume process where the animal remains intact from harvest to the consumer's table (i.e. live lobsters or whole fish), simple bands or tags with either handwritten or machine- generated tracking data have been utilized to successfully provide information on the source and age of the animal. The tracking of shellfish, generally accomplished through this kind of tag, was one of the first applications of traceability. Likewise, processors have found that utilizing handwritten tags, cards or records provide the ability to capture and retain important trace data which is then associated with the "lot" or "batch" of process output such as finished products or work-in-process (WIP). These outputs are generally marked with lot numbers which then allow receiving parties and users to identify these discreet units. In some cases, receivers will assign their own lot numbers to meet internal tracing needs or in cases where no lot number has been utilized by the upstream trading partner.

Historical methods of maintaining trace data in handwritten records create significant impediments to effective and efficient trace activity. Conducting trace forward and trace backward exercises are inherently slow and subject to error, especially in larger, complex supply chains. Handwritten records make it much more difficult to conduct analysis to support business improvement and to proactively identify issues or trends. Handwritten and manual filing systems make it much more difficult to associate additional information to product lots beyond simple source and age information. Without these additional associated data, businesses are finding that they are challenged to provide the type of transparency that regulators, customers and consumers are demanding today.

Given the diversity of species, products and supply chains that make up the seafood industry, there is a multitude of approaches to employing technology to meet increasing business, regulatory, customer and consumer tracing needs. Many businesses, especially medium to large businesses, are utilizing best in breed Enterprise Resource Planning (ERP) systems to collect, store and share traceability data. Excel, Access, and other database programs are being utilized effectively in less complex businesses. In recent years, there has been a proliferation of third party technology providers supplying end to end trace tools to the seafood industry. Integration of coding symbology, RIFD tags, GPS devices and extensive use of the internet is enabling progressive seafood businesses to effect traceability in real time and to provide a level of transparency in product data to provide important points of differentiation in the marketplace. Some seafood businesses are offering tools to consumers which allow them to enter coding information, scan codes or take pictures of products and immediately get detailed product information, including its sourcing on their computer or smartphone.

Traceability as described above is a mechanism supporting business improvement and especially in monitoring food safety to trace back to the source(s) of adulterants in food and tracking forward to get misbranded, mislabeled, or unsafe food off the shelves. As in all food industries, food safety in seafood is extremely important, but there are other applications of traceability that help leverage the cost of trace programs. For example, business benefits of traceability that were reported during the FDA Pilot Study included shrinkage cost savings of $3000 per week, a new business relationship (valued at $4 million) because of the supplier's product traceability plan, and benefits of $200,000 from improved supply chain management [12]. Other applications in fisheries management are very much current topics in seafood as follows:

Sustainability

The NOAA Fish Watch Program defines sustainable seafood as "catching or farming seafood responsibly, with consideration for the long-term health of the environment and the livelihood of the people that depend upon the environment." Verifying the health and sustainability of U.S. and international fisheries is not always simple.

Domestic fisheries are managed by State and Federal agencies under legally established fisheries management plans. International fisheries are managed under sovereign laws and international treaties. Guidance on how to make sustainable seafood choices is found on the NOAA FishWatch site [8].

Meeting the ever-growing demand for seafood around the world is an opportunity for the global seafood community. Aquaculture is central to our ability to meet the seafood demands of an ever increasing world population. The Global Aquaculture Alliance has duly noted "the aquaculture gap," meaning aquaculture product will need to increase by 46.4 million metric tons by 2030 to meet the world's demand for seafood [9].

The subject of sustainability is integral to seafood as a commodity for healthy eating, and the industry is most interested in ensuring fish in commerce is sustainably caught. One way to do that is through traceability. Other parties such as National Oceanic and Atmospheric Administration (NOAA), non-government organizations (NGOs), third party certifiers, technical solution providers, and retailers are stakeholders in traceability, and so the topic is currently a lively discussion point. Basic KDEs of sustainability recommended by the National Fisheries Institute (NFI) are the Latin names of the species, its method of production, and the geographical region of production. (Fig. 8.1) An additional data element in production methods of wild catch includes gear types which is an element in discussion among the industry and seafood NGOs as to its importance in KDEs. Gear types are

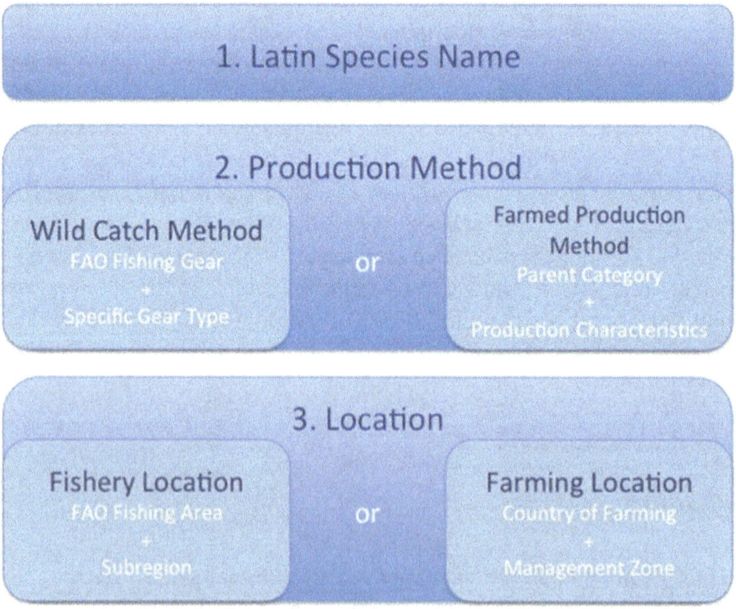

Fig. 8.1 Key data elements for sustainability recommended by the National Fisheries Institute. Note sub-categories of supporting information

classified based on the FAO Fisheries Technical Paper 222 [4] titled "Definition and classification of fishing gear categories, T222, Rev.1." Fishing gear types frequently include a "parent" list with sub-categories such as beach and boat seines under the category of seine nets [4]. Additional data elements, and optional in NFI's view, for farmed production methods are classified by system and environmental impact in FAO's online document "Aquaculture Methods and Practices: A Selected Review [2]." Fishery location is by the FAO numerical designation of major fishing areas of the world [5] with any applicable subareas and latitudes/longitudes and one of the following: (1) country of harvest if caught in the Exclusive Economic Zone (EEZ) of that country, (2) by the Regional Fisheries Management Organization (RFMO) name if that applies, or (3) by the high seas or ocean designation. Farming location is the country where the farm is located. Other KDEs besides Latin name, production method, and location may be demanded by retailers (e.g., certification by third party auditors, ranking for sustainability by NGOs), though these are individual requirements between supplier and retailer. No regulatory requirements have yet been established.

Certification of sustainable catch complements the advice of the 2015 U.S. Department of Agriculture (USDA) Dietary Guidelines for Americans (DGA) Committee (DGAC) which, like the 2010 DGA Committee, recommended increased consumption of seafood but also considered sustainability. If Americans are to double the average intake of EPA (eicosapentaenoic acid) and DHA (docosahexaenoic acid) fatty acids per day, they must double their per capita fish consumption [14], and that leads to managing the availability of seafood. The DGAC recognized that aquaculture and its management practices are becoming increasingly important in providing seafood to the world, but DGAC expressed concern that aquaculture may expand to those species with lower amounts of DHA and EPA. Perceived exploitation of ocean fisheries and consideration of eco-system impact has prompted many third-party certifiers (e.g., Alaska Responsible Fisheries Management, Marine Stewardship Council, Monterey Bay Aquarium) to define their criteria by certifying which seafood stocks are at sustainable levels and offering eco-labels, websites, quick reference guides, etc. to communicate that message to consumers. Sustainability is a must for certification. Certification is a must for eco-labeling, but neither guarantees sustainability.

Since 1976 the Magnuson-Stevens Fishery Conservation and Management Act (MSA) has been the primary statute governing U.S. fisheries (National Marine Fisheries Service/NOAA). The National Oceanic and Atmospheric Administration (NOAA) is the federal agency that implements the Act. NOAA is responsible for the management, conservation, and protection of living marine resources within the U.S. EEZ which is defined as waters three to 200 miles offshore.

Illegal, Unreported, and Unregulated (IUU) Fishing and Seafood Fraud

IUU

Illegal fishing consists of fishing activities conducted in contravention of applicable laws and regulations, including those laws and rules adopted at the regional and international level. Unreported fishing refers to those fishing activities that are not reported or are misreported to relevant authorities in contravention of national laws and regulations or reporting procedures of a relevant RFMO. Unregulated fishing occurs in areas or for fish stocks for which there are no applicable conservation or management measures and where such fishing activities are conducted in a manner inconsistent with State responsibilities for the conservation of living marine resources under international law. Fishing activities are also unregulated when occurring in an RFMO-managed area and conducted by vessels without nationality, or by those flying a flag of a State or fishing entity that is not party to the RFMO in a manner that is consistent with the conservation measure of that RFMO.

In June 2014 the Department of State hosted a conference titled "Our Oceans." One of the three pillar topics was sustainable fisheries. On Day 2 of the conference a Presidential Memorandum was issued to "Establish a Framework to Combat IUU Fishing and Seafood Fraud." In the Memorandum President Obama stated, "… IUU fishing continues to undermine the economic and environmental sustainability of fisheries and fish stocks, both in the United States and around the world. Global losses attributable to the black market from IUU fishing are estimated to be $10–23 billion annually [1], weakening profitability for legally caught seafood, fueling illegal trafficking operations, and undermining economic opportunity for legitimate fishermen in the United States and around the world." The Memo further stated that all executive departments and agencies would address these issues by improving transparency and traceability in the supply chain. The Task Force subsequently established met over the ensuing months to make 15 recommendations to the President that were published in the Federal Register on December 18, 2014 [18]. The Task Force is now the National Ocean Council Committee on IUU Fishing and Seafood Fraud and is co-chaired by NOAA and the Department of State. The work of this committee led to regulations effective January 9, 2017 and issued by NOAA that established the Seafood Import Monitoring Program [15], a risk-based traceability program, which applies to Atlantic and Pacific cod, blue crab, mahi mahi, grouper, king crab, sea cucumber, red snapper, sharks, swordfish, and tuna. As of December 31, 2018 shrimp and abalone were included in this list. The intent of the program is to verify lawful harvest of the designated wild caught and aquaculture species through chain of custody data collected in the government database, the International Trade Data System. The importer of record must provide key data from point of harvest to point of entry into U.S. commerce. The importer of record must retain its records for 2 years.

Seafood Fraud Prevention

Seafood fraud undermines the economic viability of U.S. and global fisheries and deceives consumers about their purchasing choices. Such fraud can occur at multiple points along the supply chain and includes such practices as species substitution or improper net weight declaration in which glaze weight or added water weight is inaccurately reported on the label. Seafood may be intentionally transshipped to avoid anti-dumping and countervailing duties, thus mislabeling for the country of origin. Seafood fraud and IUU fishing can overlap when there is mislabeling or other types of deceptive marketing with respect to origin or species.

FDA has established a compliance program for identifying seafood for correct species name. The multi-pronged Fish SCALE (Seafood Compliance and Labeling Enforcement) program includes the development of validated DNA testing methods along with a library of DNA sequence data for species which have been authenticated with taxonomically identified specimens and sampling assignments to pull samples from imports, warehouses, distribution centers and retail. FDA has conducted limited testing from the distribution chain prior to retail and found the rate of mislabeling to be 15% [7]. Grouper and snapper were the species most at risk for mislabeling. A published investigation [10] of mislabeling in the restaurants and at retail in one of three U.S. cities (Austin, TX, New York City, and San Francisco) has shown the rate to be 12.8%. Interestingly, the rate of mislabeling at restaurants in San Francisco was 14.8% and at retail was 2.2%. Snapper, including red, and basa were top species mislabeled in this study.

FDA has taken compliance actions such as Warning Letters, injunction orders, and Import Alerts against seafood firms for misbranding violations determined with DNA testing. The Action Plan of the President's Task Force on IUU calls for traceability on species at risk for mislabeling because such practices are a means to avoid detection of illegally caught fish [18]. At this writing NOAA is soliciting comment on which are the high-risk species.

Seafood Industry Initiatives in Traceability

In late 2009 the National Fisheries Institute made a decision to prepare a platform for traceability. A Task Force was commissioned among the membership, and the process began to write a "how to" guide on seafood traceability based on discussions with the Institute of Food Technologists (Chicago, IL), which was and still is a known authority in traceability, and GSI US, the global traceability standards group. Titled Traceability for Seafood/U.S. Implementation Guide ("Seafood Trace Guide"), this document was first released at the 2011 Boston Seafood Show as a voluntary guide for all the seafood industry [13].

Requisite to designing any traceability system is understanding the flow of the product. In the case of seafood, that means water to table. Figure 8.2 illustrates the

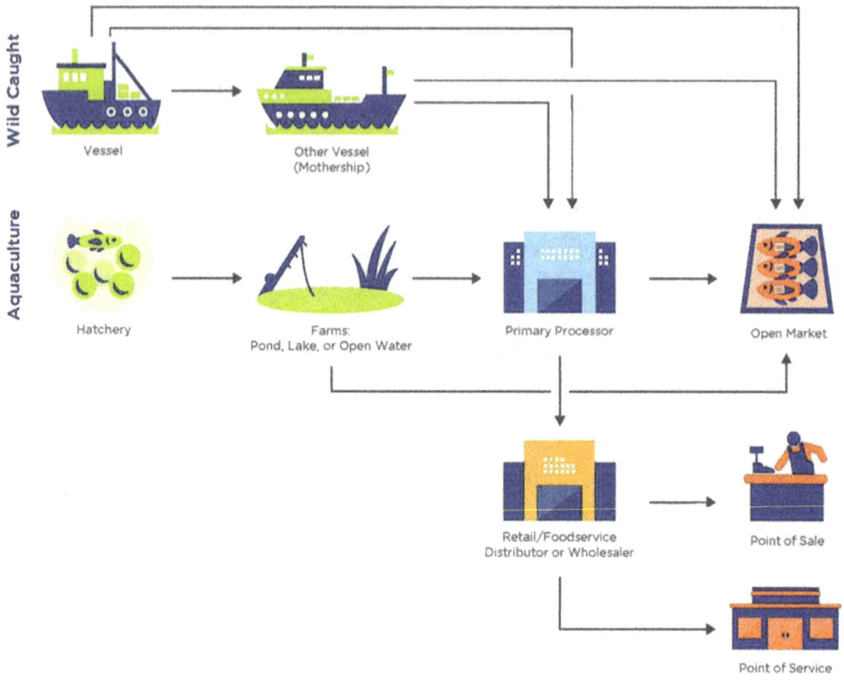

Fig. 8.2 An overview of the seafood distribution supply chain and the roles played by various distribution partners [13]

seafood supply chain, whether the species is farmed or wild caught, which is not always a linear chain. Especially for wild caught, fish may be captured by smaller boats that take their load to a mother ship. The fish may then be processed on the ship or in a plant, sold whole, headed and gutted, filleted, or value added, and shipped to a distribution center or wholesaler before going onto point of sale, either retail or food service.

When the Seafood Trace Guide was prepared, industry writers quickly recognized the importance of unique identification of product and recommended all the seafood industry use GS1 standards for tracking. However, the group recognized that there are smaller companies that do not need a formalized tracking system. These can use paper-based systems. The more complex the supply chain, the more efficiency can be derived with the use of GS1 standards in monitoring flow.

The importance of traceability systems was particularly noted in 2012 when FDA was faced with responding to a multi-state outbreak of *Salmonella* Bareilly and *Salmonella* Nchanga associated with Nakaochi scrape (frozen raw yellowfin tuna) made by Moon Fishery in India. The recall originally involved 58,828 pounds of tuna. While there were no deaths, over 425 people in 28 states were sickened. At least 55 were hospitalized [3]. FDA spent months in product tracing challenges hampered by:

- Lack of rapid connectivity
- Lack of a unique identifier
- Repacking and co-mingling
- Proper addresses and shipping and receiving dates
- Legibility and accuracy of records
- Packaging no longer available
- Product no longer available
- Delays in providing records

The challenges above were not unique to a seafood recall. While the Seafood Trace Guide was prepared before this major recall, all food manufacturers are well advised to note the challenges as addressing them will make FDA's Coordinated Outbreak Response and Evaluation (CORE) Team's job more easily accomplished. Because of this recall, the agency noted the three key areas for improvement are product connectivity/linkages, documentation, and speed. The Seafood Trace Guide focuses on these key areas.

Figure 8.3 offers a detailed diagrammatic view (on the left side) of the supply chain in terms of CTEs (e.g., event data, product data, and transport data) and, on the right, KDEs to support those CTEs. Note the use of GS1 standards. Best practices dictate at least one information source be listed on the product:

Global Trade Identification Number (GTIN) or some form of item identification;
Global Location Number (GLN) or some form of the traceability partners' name and address;
GLN of the physical location for the targeted product;
Dates or time periods for the targeted product;
Lot number

Such an approach is not new. The Bioterrorism Act of 2002 previously discussed requires that distribution channel participants collect, record, store, and share minimum pieces of information for traceability. More aggressive in scope than the Bioterrorism Act, traceability programs such as detailed in the Seafood Trace Guide recommend a far more granular approach to information (i.e. KDEs) recorded at each transition point (i.e. CTEs) in manufacture from water to point of sale. The Guide recommends that all items be traced forward or backward using global and unique identification, and all items in distribution should be subject to internal and external traceability to ensure linkages between input and output [13]. The Bioterrorism Act is not directed at internal linkages.

The spreadsheet shown in Fig. 8.4 is applicable to any food product and quite adaptable to seafood. Data acquisition is given at three levels: (1) required, (2) best practice, and (3) conditional depending on business circumstances. On the spreadsheet CTEs are generic, allowing only for one transportation event, one transformation event, and one depletion event, though in practice these events may be multiple. Recall that as with sustainability, there are yet no government guidance documents or rules on CTEs and KDEs.

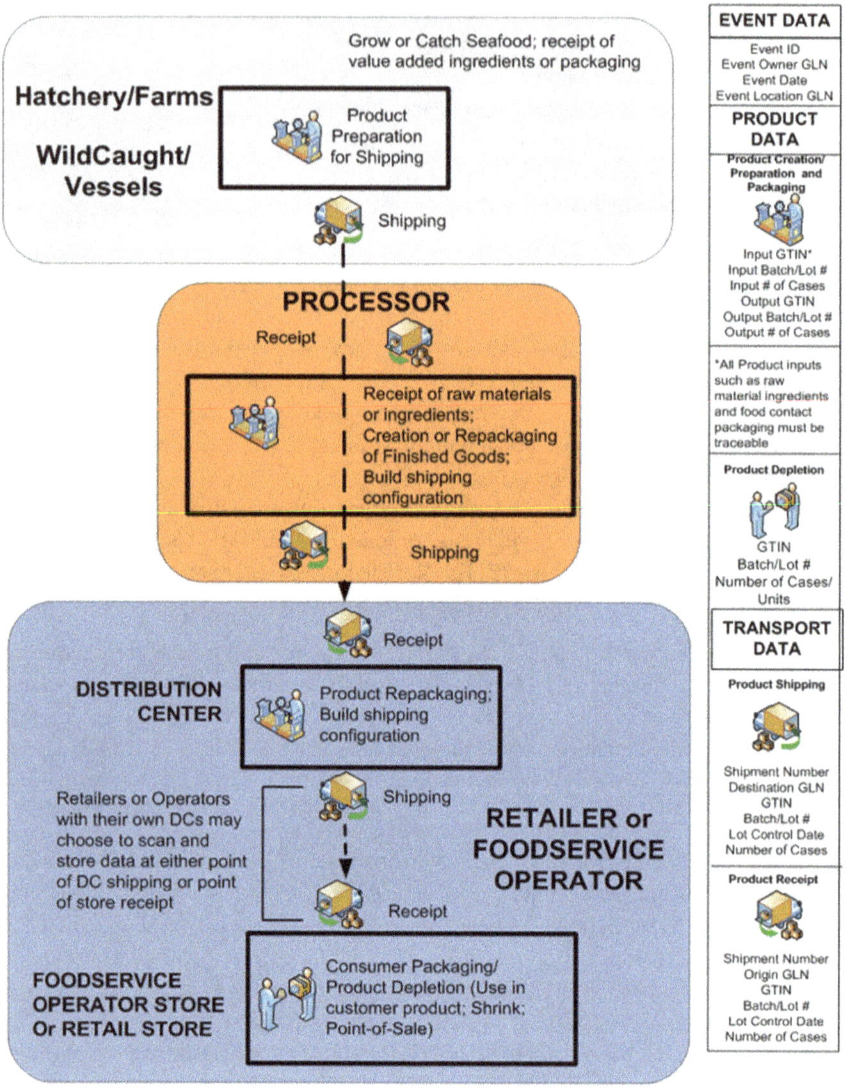

Fig. 8.3 CTEs in seafood traceability are shown in the left-hand portion. The right vertical strip shows GS1 standard key data elements to support the CTEs [13]

Seafood industry members continue to show increased interest in enhancing seafood traceability and are willing to make investments in technology to achieve this goal, but adoption is lagging in some sectors of the industry due to a variety of concerns. Competition can be fierce in the seafood industry and this makes collaboration more difficult. The seafood industry is made up of many smaller businesses that may not have the time or resources to focus upon these types of initiatives. Additionally, the global nature of the seafood supply chains means an added degree of difficulty while navigating different regulatory, cultural and market needs.

Key Data Element	Transport		Transformation		Depletion	
	Shipping	Receiving	Input	Output	Consumption	Disposal
Event Type	R	R	R	R	R	R
Event Owner	R	R	R	R	R	R
Date	R	R	R	R	R	R
Time	R	R	R	R	R	R
Event Location	R	R	R	R	R	R
Item ID Type	R	R	R	R	R	R
Item ID	R	R	R	R	R	R
Batch/Lot/Serial#	BP*	BP	R	R	BP	BP
Quantity	R	R	R	R	R	R
Unit of Measure	R	R	R	R	R	R
Batch/Lot Relevant Date	C^	C	C	C^	BP	BP
Activity Type	C	C	R	R		
Activity ID	C	C	R	R		
Supplier Identity	C	C	C	C		
Trading Partner Location	R	R				
R = Required Data						
C = Conditional Data; The need for this data would be determined by business circumstances; ^ Relevant Date should be reported by Suppliers for Shipping Events and for Transformation Output events.						
BP = Best practice is to capture the batch/lot number for transport and depletion events whenever possible; however, if not feasible, Batch/Lot Relevant date or Activity ID must be provided. * Batch/lot/serial numbers should be reported by Suppliers for Shipping events.						

Fig. 8.4 Spreadsheet of KDEs at points of CTEs as recommended by the Institute of Food Technologists in its report on pilot projects for the Food and Drug Administration [12]

While there is growing interest in adoption of GS1 standards in Food Service and Retail, employment of those standards remains at only 50% industry-wide and even lower in the seafood sector. Additionally, many industry players are waiting to understand how FDA's FSMA traceability regulations will impact their current systems and future plans. Some businesses remain concerned about providing transparency to supply chain information or product information for fear of losing a competitive manufacturing advantage or for fear that product information could be used in an unfavorable way. Concern exists that bringing transparency to extended supply chains, while working to enhance vessel registration systems and Port State Measures to combat IUU fishing, might be a case of getting the cart before the horse. For many businesses this translates into a concern about ROI on trace system investments when these issues have yet to be resolved.

Despite these challenges, the seafood industry should continue to work to remove these barriers and lead the broader retail and food service traceability initiatives, as its diversity in products, processes and markets uniquely qualify it to successfully inform these enhancement efforts. Ultimately, the consumer and customer will dictate market needs that will translate into industry actions to enhance the speed, accuracy, quantity and transparency of product tracing data and information. Being a leader in these efforts are in the seafood industry's best interests.

For complete details on preparing a traceability plan and implementing GS1 standards for seafood traceability, the reader is referred to Seafood Trace Guide available online.

Future of Seafood Traceability

Seafood is poised to lead the way in traceability in the food industry because the commodity is packed and sold shelf stable in the center of the store, in the frozen food aisle, and fresh at the perimeter of the store. Both static (e.g., species identity) and dynamic (e.g., vessel, gear type) data are captured to support traceability offering further examples to the food industry of managing various types of data. Seafood is a globally traded commodity which means if any food must strive to achieve international harmonization, it is seafood. GS1 standards, used by about 50% of the seafood industry, offer a mechanism for data capture and are well-established in international trade. Additionally, many companies and consumer groups are uniting to pilot block chain technology in an effort to utilize distributed ledgers, once used exclusively in the financial industry to manage financial transactions, to facilitate whole chain traceability. Should this technology prove out, it may be possible not only to meet government regulatory needs, but those of consumers for transparency of information. Imagine a world in which each plate or package of food would come with information on the source of where it was caught, where it was processed, where it was imported and who was responsible for each step in the process from ocean to plate. As the federal government puts additional requirements for seafood traceability in place, the importance of GS1 standards will increase even more. If the seafood industry is to be transparent throughout the supply chain, that transparency requires interoperability, the ability to exchange product tracing information accurately, efficiently, and consistently. Suppliers, processors, and customers all have "islands" of data they offer or require, all of which must have bridges of communication. How to build those bridges is the next challenge in seafood traceability, and the process is just beginning under the leadership of the Global Food Traceability Center, a part of the Institute of Food Technologists (Washington, D.C.), in which the National Fisheries Institute has an important role.

References

1. Agnew D, Pearce J, Pramod G, Peatman T, Watson R, Beddington J, Pitcher T (2009) Estimating the worldwide extent of illegal fishing. PLoS One 4:e4570
2. Baluyut E (1989) Aquaculture systems and practices: a selected review. Rome (Italy): United Nations Development Programme. 89(43) of ADCP Report
3. CDC – Multistate outbreak of salmonella bareilly and salmonella nchanga infections associated with a raw scraped ground tuna product (final update) [Internet]. Atlanta; Centers for Disease Control and Prevention; c2012 [updated 2012 July 26; cited 2015 March 2]. Available from: http://www.cdc.gov/salmonella/bareilly-04-12/advice-consumers.html
4. FAO Fisheries technical paper. No.222. Revision 1. Rome, FAO. 1990. 92p
5. FAO Fisheries & aquaculture – FAO major fishing areas [Internet]. [Place unknown]; Food and Agriculture Organization of the United Nations; c2015 [cited 2015 March 3]. Available from: http://www.fao.org/fishery/area/search/en

6. FDA Food Safety and Modernization Act of 2011, Pub. L. No. 111–353, 3885 STAT (January 4, 2011)
7. FDA DNA testing at wholesale level to evaluate proper labeling of seafood species [Internet]. [Place unknown]: FDA; c2015 [cited 2015 Sep 3]. Available from: http://www.fda.gov/Food/GuidanceRegulation/GuidanceDocumentsRegulatoryInformation/Seafood/ucm419982.htm
8. FishWatch: Choosing sustainable [Internet]. [Place unknown] NOAA; c2015 [cited 2015 Sep 4]. Available from: http://www.fishwatch.gov/buying_seafood/choosing_sustainable.htm
9. Global Aquaculture Alliance. Why It Matters. Available from: https://www.aquaculturealliance.org/what-we-do/why-it-matters/
10. Khaksar R, Carlson T, Schaffner DW et al (2015) Unmasking seafood mislabeling in U.S. markets: DNA barcoding as a unique technology for food authentication and quality control. Food Control 56:71–76. https://doi.org/10.1016/j.foodcont.2015.03.007
11. Mandatory Country of Origin Labeling of Fish and Shellfish, 7 C. F. R. Part 60 (2004)
12. McEntire J, Bhatt T Pilot projects for improving product tracing along the food supply system-final report [Internet]. Chicago: Institute of Food Technologists; c2013 [cited 2015 Sep 4]. Available from: http://www.fda.gov/downloads/Food/GuidanceRegulation/UCM341810.pdf
13. National Fisheries Institute and GS1 U.S. traceability for seafood, U.S. implementation guide [Internet]. [Place unknown]: NFI and GS1 U.S; c2015 NFI and GS1 U.S; c2011 [cited 2015 September 4]. Available from: http://www.aboutseafood.com/about/us-seafood-traceability-implementation-guide
14. Nesheim MC, Nestle M (2014) Advice for fish consumption: challenging dilemmas. Am J Clin Nutr 99(5):973–974. https://doi.org/10.3945/ajcn.114.086488
15. National Oceanic and Atmospheric Administration/NOAA. Seafood Import Monitoring Program. Available in: https://www.fisheries.noaa.gov/national/international-affairs/seafood-import-monitoring-program
16. Procedures for the Safe and Sanitary Processing and Importing of Fish and Fishery Products, 21 CFR Parts 123 and 1240, (1995)
17. Public health security and bioterrorism preparedness and response act of 2002. Pub. L. No. 107–108, STAT. 594 (June 12, 2002)
18. Recommendations of the Presidential Task Force on Combating Illegal, Unreported and Unregulated Fishing and Seafood Fraud, 79 C. F. R. 75536–75541, (2014)

Chapter 9
A Chain of Linked Nuances

Lisa Jo Lupo

Abstract It seems relatively simple: All of the food we eat originates on the "farm" as produce or animal, then flows downstream through a watercourse of channels to arrive on the consumer's plate. Unfortunately the food supply chain is not quite that simple, nor is it always a forward flow. In fact, it can be argued that the supply chain is actually driven backward, with the demands and expectations of the consumer creating ripples that impact each link of the chain – from retail (grocery, restaurant, or farmers' market) back through distribution, manufacturing, and packing/co-packing, to the farm. And when a customer complaint or positive test result necessitates tracing back to the source and forward for recall and communication, the unique nuances and challenges, the stressors and strains, of traceability at each link – and the potential results of breaks in the chain – are found to create a rather complex torrent of channels that defy the perceptively easy flow of "downstream." Thus, while the flow of the food supply system is often referred to as upstream and downstream, seeing it as a series of links in a chain is, in fact, a better representation: each link is a separate entity but each must seamlessly interconnect with the link to each of its sides for the system to be successful as a whole. This chapter follows that chain (from the consumer backward) to discuss each link and linkage, and the nuances and challenges that are created by the riptides of back-flowing expectations and forward-flowing product/ingredient identification.

Keywords Retail · Upstream · Downstream · Customers · Manufacturing · Farm

L. J. Lupo (✉)
LJ Writing Services, Stow, Ohio, USA

Quality Assurance & Food Safety (QA), Valley View, OH, USA
e-mail: lisa@ljwriting.com

J. McEntire, A. W. Kennedy (eds.), *Food Traceability*, Food Microbiology and Food Safety, https://doi.org/10.1007/978-3-030-10902-8_9

Introduction

Today's food production and distribution system is evolving from processes of reaction to those of prevention. Traceability is, in and of itself, a reactive system – intended to trace a problem that has already occurred back to its source, but, because our world is not a perfect one, and errors and misguided intentional issues do occur, traceability – with its integrated corrective action – is an integral aspect of prevention in the food safety system at each link in the chain.

Traceability is not a new concept, but it is an ever-evolving thing, with today's traceability driven not simply by food safety, but also by consumer demand for transparency and accountability, and the use of their voices and purchasing power to impact the chain of food production.

There are similarities in the impacts, nuances and challenges of traceability between the links of the food chain – and these commonalities have increased with the publication of the rules of the Food Safety Modernization Act (FSMA), intensifying the equitability and responsibility for food safety along the chain by adding or increasing regulation to links in the chain not previously held to such rules – such as that of transportation with its significant impact on distributors, and produce which now has its own specific set of federal rules.

Additionally, some of the generalities of challenges are of the same vein. For example, it is commonly held that tracing backward is more difficult than tracing forward. In forward-tracing, the points tend to narrow down; but backward – most finished product is either multi-ingredient, each of which must be individually traced back, or, if bulk product such as produce or grains, is likely to have been commingled at at least one point in the chain. Thus, the hand-off between links in the chain is of critical importance.

Food products are currently able to be labeled to the case level, but both the desire and growing requirements are to label to the individual product. Is it possible? The ability exists, but the economic feasibility and the attainable accuracy remain in question. Being able to individually label ever-lower amounts of product will not only help to further protect the individual consumer, it will reduce food waste as companies exercise an "abundance of caution" to dispose of not just entire lots but often one lot forward and one lot back, for the ultimate in consumer (and brand) protection. Such waste is evolving from a business concern to a global concern as the world population continues to increase to an expected 9.8 billion by the year 2050. It is such human factors that are driving the food industry to increase its accountability to the consumer – at each link in the chain.

The Consumer: An Integral Part of the Chain

Just as all food ends with the person who intends to consume it, so too do the demands and expectations of and for that food begin with that same consumer. Thus, to thoroughly understand the nuances and challenges of each link in the chain

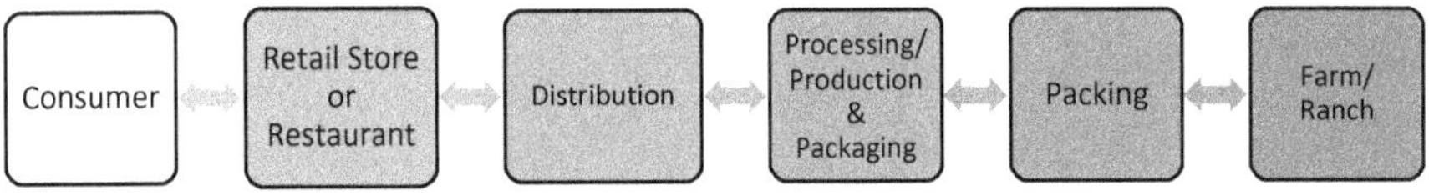

Fig. 9.1 Supply chain: The consumer

that supplies this food – from farm to table, we must consider consumers to be an integral link; understand their expectations and perceptions (right or wrong); and realize the impact of these on the demands of traceability (Fig. 9.1).

Today's consumers are more educated about the food they eat. They seek out information and have unprecedented access to information – and misinformation – through the virtually infinite expanses of the internet – the World Wide Web – and all the topics and trends therein. Because the internet and its social media options provide open, unrestricted, uncontrolled, and often anonymous, access to anyone who wishes to opine on anything, those who have anything to say – fact or fiction – have unprecedented access to a worldwide platform and a receptive audience.

There's an old saying that a person will tell of a bad experience to ten people who will tell ten people and so on, eventually reaching hundreds of people. In today's world with the ability of a single post to be liked, shared, tweeted, retweeted, discussed, snapchatted, etc., that saying is multiplied a hundredfold and more, with the ability of a single person's idea or opinion to become an overnight trend, which soon transitions to a consumer demand, and, more often than not, is developed as a retail standard or even federal regulation.

One simply need recall the impact of a mother's post on "pink slime" in 2012 to understand the impact. Although lean finely textured beef (LFTB) was approved by the USDA as safe and used by numerous ground beef retailers, its internet sensation had consumers in an uproar, and led to school-lunch bans, the closing of three of the company's four plants, and loss of hundreds of jobs. Ironically, barely a year later, economic concerns and the USDA's continued affirmation of the product's safety put the beef product back in schools and on consumers' tables – and even holds a place of prominence on its manufacturer's website.

Like the perception of the LFTB … many consumer trends have little or nothing to do with food safety – despite such misrepresentation and misinformation that is all too prominent on the Web.

And it is just such trends and misinformation that add transparency to the list of challenges the food industry faces in traceability. Consumers want safe food – which is the key reason for traceability, but they also want to be able to make informed purchasing decisions, and retailers want to provide the means for them to do so.

Thus, traceability has become more than a tracking of contaminants, adulterants, and unlabeled ingredients. It has become a means of fulfilling consumers' desire, and right, to know what is – and isn't – in their food; how their food is grown …

manufactured … stored … transported … served; what really happens behind the closed doors of the food supply chain.

Walmart: Dedication to Transparency

Oct. 6, 2014 – "In front of hundreds of associates, suppliers and nonprofit organizations at its Global Sustainability Milestone Meeting, Walmart today announced its commitment to create a more sustainable food system. The company will reach this goal through four key pillars: improving the affordability of food for both customers and the environment, increasing access to food, making healthier eating easier, and improving the safety and transparency of the food chain. … Walmart will work to provide more information and transparency about the products on its shelves so customers can see where an item came from, how it was made, and decode the ingredient label." [8]

May 22, 2015 – "Our customers want to know more about how their food is grown and raised, and where it comes from. As the nation's largest grocer, Walmart is committed to using our strengths to drive transparency and improvement across the supply chain," said [Senior Vice President of Sustainability Kathleen] McLaughlin. "We believe it's important to promote transparency in this process, helping to put our customers in charge of their food choices by providing clear, accurate information about food ingredients. We appreciate the leadership our suppliers have shown to help us accomplish these goals." [10]

Thus, traceability has taken on an expanded role in each link in the chain. Not only must retailers ensure that the produce in their bins is safe, that allergen-free food truly is allergen free, but the supplier who provides that retailer and foodservice company with food labeled as organic, natural, or GMO-free is being driven to follow supplier traceability standards and recordkeeping of such transparency. The same is true of such consumer demands for humane treatment of animals, antibiotic-free, pesticide-free, cage-free, etc.

Disregarding one's opinion on the truth or misrepresentation of any of these, consumer right to know has added a challenging dimension to traceability. It is a matter of brand protection of each link to ensure every ingredient in every product is traced back to its root source, with each and every label claim – regardless of food safety or quality applicability – validated each step of the way.

It's a matter of consumer confidence in and future purchase of a brand's products. As depicted by the results of the 2018 Food and Health Survey from the International Food Information Council Foundation, consumers continue to be concerned about foodborne illness, carcinogens and chemicals in foods, but confidence in the overall food supply has risen slightly with the increased regulation (Table 9.1).

Table 9.1 Consumer confidence in the food supply [1]

Confidence level	2014	2018
Very/somewhat	66%	68%
Not too/not at all	30%	28%

Even the regulatory aspects of the food supply chain are essentially driven by the consumer. One simply need look at the Food Safety Modernization Act (FSMA) to understand this. While the specific rules of FSMA are written and enforced by the U.S. Food and Drug Administration (FDA), the Act itself, which requires FDA publication of the rules, was written, passed and mandated by Congress – an elected body "of the people."

And according to the 2015 Harris Poll from Nielsen, "the people" believe such government oversight is critical. Of poll respondents:

- 86% say food recalls have them at least somewhat concerned (with 58% somewhat concerned and 28% seriously concerned).
- 73% believe there should be more government oversight in regard to food safety.

When all this is taken into consideration, it is fairly easy to see that consumers drive retailers; and in order to meet consumer demands and expectations, those retailers (grocery and foodservice) must not only drive their suppliers, driving those suppliers' suppliers and the suppliers' suppliers' supplier … back to the farm, they must have a traceability system that verifies and validates the food safety [5].

At Retail: A Dual Role of Grocery and Foodservice

One of the greatest challenges for those in the commercial marketplace – whether it be retail or restaurant – is its dual role. Not only is it constrained by its own consumer, regulatory, and corporate requirements, but because it has the most direct line to consumer purchase, it is the link which has the responsibility of removing unsafe or mislabeled products from consumer purchase (Fig. 9.2).

Additionally, it is generally the only link that has any potential for tracking such product *after* consumer purchase. With such increased accountability, those who sell direct to consumer have learned to take on a policing role in traceability.

As such, anything that is done at this level is done for (or to) everyone in the supply chain. A restaurant that decides to promote its "all-natural, additive-free" menu has an obligation to hold its suppliers, its suppliers' suppliers, etc., to that commitment as well.

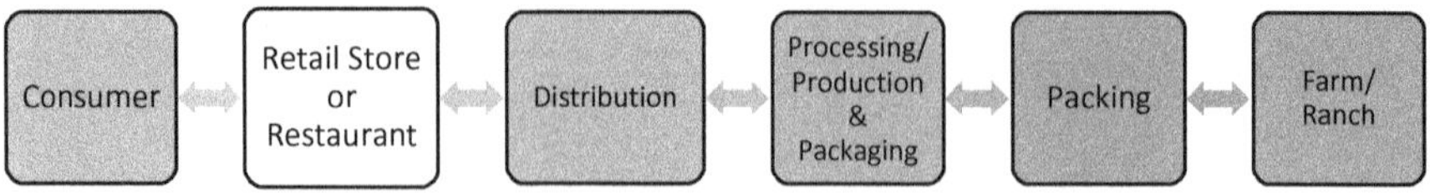

Fig. 9.2 Supply chain: Retail store and restaurant

Restaurant and Retail Impact

Every restaurant and retailer, large or small, impacts the food chain back to its source, even if only in its decision as to which products it will carry. But, there is no question that the larger and more prominent it is, the greater impact it will have. One of the best demonstrations of this is that of Walmart Inc.

With revenues of more than $500 million, Walmart was No. 1 on the 2018 Fortune 500 list [11]. Whether one has a positive or negative view of this corporate giant (and/or corporate giants as a whole), Walmart's prominence has enabled it to lead the food chain to increased food safety, transparency, and traceability.

As then-Vice President of Food Safety Frank Yiannas said in an interview with Quality Assurance & Food Safety (QA) magazine, "140 million customers walk through our stores in the U.S. in one week, and 200 million do globally … we have the ability to have a large impact on food safety and health." [9]

Those customers, Yiannas said, have an unspoken expectation that the products they buy will be safe. And ensuring that the expectation is fulfilled means requiring a culture of food safety throughout its stores, *and* throughout the company's entire supply chain.

While each step back in the supply chain has customer commitments for which it must make demands of its suppliers (e.g., an organic processor must ensure its suppliers provide only verified-organic ingredients, a peanut-free product manufacturer must ensure no cross-contact back to the farm, etc.), the consumer-facing company has the longest and most complex of chains to trace and the greatest obligation to quickly know of recalls, and stop sales of implicated product.

The Bioterrorism Act of 2002 recordkeeping requirements mandate that every link in the chain be able to trace products one step forward and one back, with lot number as the standard identifier. Although there are some retail and restaurant exemptions to the mandate, as well as new FSMA recordkeeping requirements, in today's litigious world, it behooves restaurants and retailers to consider themselves as much an integral link in the traceability chain as they are in the food chain itself.

It is, however, the very identification of implicated product that can be the greatest challenge in traceability for the grocer or foodservice provider, particularly when dealing with loose produce item, such as apples, onions or tomatoes.

In order to ensure a continuous and abundant supply of produce for customers throughout the day, items from varying lots will be commingled in the retail produce bins as stock is continually replenished, and/or product is separated for use in deli sandwiches or salads. While the produce is still trackable by lot, as required, this mixing and secondary use of products of multiple lots means that the recall of a single lot of tomatoes could cause a store to have to pull every tomato in the store bins, deli prep, and prepared tomato sandwich and salad.

Thus, depending on the commodity, how much is out in the store at the time, and how the supplier defines its "lot," a recall of a produce item could mean a retailer pulls 20 or 200 pounds of product per store – which may or may not be of the lot recalled.

That said, retailers such as Walmart, require that every lot be traceable back through production and/or packing house (facility, date, time, line number, etc.) to the farm (field, date, time, etc.). The information need not all be listed on the lot coding, but it needs to be traceable … in case it needs to be traced. And the faster the retailer is informed of an issue, the better for everyone. Pulling an implicated product while still at the retailer's distribution center will be much more efficient and able to be lot specific.

Even non-produce items can be subject to such an "abundance of caution," as the definitions of packaged product lots can also be subjective, with products labeled to the case or pallet level. With such items, one may choose to pull or recall not only the implicated lot but also those of previous and successive lots as well. While the ability to label to the individual product exists, and would be of great benefit in waste reduction, the economic viability is questionable.

Additionally, few retailers would have the ability to accurately inform individual customers of an implicated product they purchased. And those that do keep track of their customers' purchases through store cards or other means may be wary of beginning such a notification process because of the liability of responsibility to inform that could go along with it.

Thus, while both retailers and restaurants seek to balance food safety and traceability/recall requirements with sustainability and environmentalism, the current standard tends to focus on an "abundance of caution."

Through Distribution: The Middleman

As the next-in-line upstream supplier to the grocery/foodservice provider, distributors are faced with responding to numerous challenges to meet the expectations of the end consumer as well as fall within the standards of the retailers to whom they distribute product. At the same time, they are continually being challenged by the range of capabilities, and the vast number, of suppliers that need to be managed (Fig. 9.3).

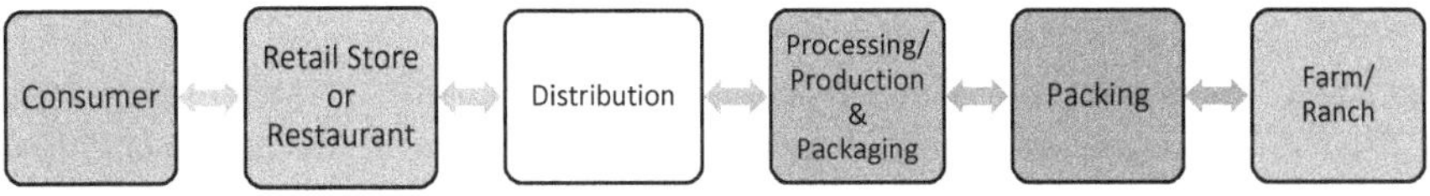

Fig. 9.3 Supply chain: distribution

As such, retailers and restaurants do not always have a true sense of the capabilities of distributors, the feasibility of certain expectations they have of this link in the chain, and/or the challenges of meeting expectations through a long-term solution rather than a short-term bandage.

Upstream Suppliers

To a great extent, the larger the distributor, the more in-depth – and more complex – its traceability program. With the need to manage more than 100,000 suppliers of varying types and sizes, a large distributor, such as US Foods or Sysco, has to be able to depend on complete and accurate information from its suppliers. However, there are as many reasons as there are suppliers that this isn't always the case.

A primary reason is the variation in lot coding. While food is required to be traceable through a lot coding system, there can be a great deal of variation between the systems and resulting lot codes of the products. For example, a produce supplier may define a lot by date and/or time, grower or field, or even buyer, and a co-packer doing small runs may simply break lots by brand – each of which would create a difference in coding. (To see the variation that exists between products, one simply need look at the array of lot codes on the boxes and cans of food in their pantry.)

Because of all this, product codes don't always align with the information needed by the distributor or that which is expected by the retailer. This may simply be due to the system the manufacturer chooses to use, or it may be a factor of the company's capabilities, with small and local suppliers often having more basic, even manual, systems. (The same can be true of distributors themselves, with smaller distributors not always having the resources or the technology of larger, multi-facility businesses.)

This, of course, in no way absolves a distributor from ensuring it can trace all product one forward and one back as required. Thus, most develop specific supplier requirements and systems to verify the supplied information. As discussed in a 2010 QA magazine profile on US Foods [3], for example, the company was tracking incoming supplies by pallet based on the receiving date and recording the issuance of pallets to customers. The specific case and lot numbers of product on each pallet were provided by the supplier on an Advance Shipping Notice or the purchase order. When shipments were received, they were spot checked for validation.

While it would be ideal for all in the supply chain to be able to apply technology that would scan every case and item in and out, the available technology is generally too expensive and time consuming for universal implementation. Thus systems still often focus on the pallet and can require significant manual input.

This is because, although electronic data capture systems are used in distribution, they are, for the most part, static. That is, general information can be pulled into the system (XYZ Beef case of beef patties, lot #150709435, etc.), but additional information that is needed from the supplier may not be able to automatically transfer to the next system, so that data, such as lot testing date, expiration date, etc., must be added as transactional data instead. With technology accelerating at a rapid rate, new systems such as blockchain and those that utilize secure internet or "cloud" sites to provide data links are being developed and tested regularly. But down-to-the-item tracking is simply not – yet – feasible for all in the food chain, particularly for small and very small businesses, those bringing in supplies from a vast range of manufacturers, and those still tracking on paper or spreadsheets.

The recordkeeping and tracking get further complex when a distributor allows broken shipments, e.g., customer purchase of less than a full case. Because a product lot could be made up of only a few cases from a small supplier or thousands of items by a large manufacturer, a recall can mean identification and notification of a single customer or that of numerous recipients, and breaking a shipment by allowing purchase of a single bulk can or a few boxes of product by many would significantly increase the tracking and notification needed. Thus, some distributors no longer allow this practice, requiring minimums and full-case purchase.

Downstream Customers

As the previous example shows, the integration of traceability systems isn't just an upstream supplier issue, rather, as the middleman of the food supply chain, the distributor may have significant challenges in tracking downstream as well. A single box or can of food, for example, may travel a number of different ways/routes before reaching the end consumer. Additionally, retailers are not the only businesses to which distributors ship. Rather, some customers receive product for further processing, such as pizza, salads, etc. This means that a single item that is sold to a consumer may be an amalgamation of numerous products – with the burden placed on the distributor to capture and validate that information back to its source.

In many cases, a distributor is actually delivering to a retailer's distribution center (DC) – adding an additional link in the chain and, generally, additional requirements. Walmart, for example, receives product at more than 150 distribution centers, from which it ships product to its retail stores. The retailer requires not only that incoming goods be tracked back to the source, but that it be informed of any other recipients of the product from the same lot.

This is because the retailer may have purchased additional supplies of the product, say apples from a broker – to whom the distributor provided the product. So by having the information in a single record, the retailer would immediately know to pull that product as well, instead of having to wait for secondary contact from the broker. And when a Walmart DC is informed of a recall of product it received, the company will tell all its stores served by that center to pull the product from their shelves – regardless of lot number – out of an "abundance of caution." The retailer also has the ability to restrict the sale of an item at the cash register, so if any shoppers put the product in their carts before it got pulled, or a single package of the recalled product was missed, the cashier will not be able to ring it up/allow purchase. Walmart doesn't lift the restriction until it is informed of and is comfortable with corrective action taken by the supplier – FDA acceptance is not always enough.

Whether a distributor is sending product to a customer's distribution center or directly to the retail site, it can face challenges of customer requests for delivery scans. In some cases (e.g., if the supplier follows GS1 standards and labeling practices), distribution information can be automated, but a requirement to scan every item at every stop would significantly increase the time and costs of distribution, increasing the overall cost of products. For example, if a distributor's delivery stop averages 25 min; scanning each item would add a minimum of 5 min to each stop. When calculated across the hours and number of stops in a business day, the impact would be a reduced ability in number of stops per driver, increasing the number of drivers and trucks needed. Because of transportation rules limiting the number of hours a driver can be on the road in a day, the company couldn't simply allow the driver to add the extra minutes, and resulting hours, to the day.

Such on-the-road rules also increased with the roll out of FSMA, which includes new rules for transportation of food product and defines distributor facilities as "food facilities," making them subject to food facility rules to which they were not previously held.

Consumer Expectations

Although the law only mandates tracing one step forward, a distributor is just as held to consumer expectations as is the retailer. For example, while a consumer request for "local" product does not increase traceability challenges – as this is listed on the label and the retailer/foodservice provider can select items by ingredients, it can become an issue when product runs low. For example, if a customer orders produce, the distributor can generally provide it from a number of sources, substituting one source for another when supply runs low. But if the customer requires local produce, and there is none available, the customer will have to make a choice, allowing produce from another region to be shipped or not receiving any. The distributor cannot simply substitute a comparable, non-local, product. The same is true of a request for natural, GMO-free, and other such consumer requests.

Stocking local product also provides other challenges, and costs, for distributors, as these have to be separately received and handled, specifically coded and segregated. Thus, while customers often perceive that local foods should cost less and be more environmental, the opposite is actually true. For example, rather than utilizing the efficiencies of large trucks that can haul a vast array of foods, local produce is more likely to be transported across the region in small amounts in pick-up trucks that have higher consumption of gas per pound of food. Thus, the environmental impact can be significantly higher for local foods – contrary to general belief.

The expectation of the distributor's customer is driven by the expectation of the retailer's customer – the end consumer. And those expectations continue to pass upstream to the manufacturer.

In Manufacturing

According to the 2013 CDC report, *Surveillance for Foodborne Disease Outbreaks – United States, 2009–2010*, "Among the 766 [foodborne illness] outbreaks with a known single setting where food was consumed, 48% were caused by food consumed in a restaurant or deli, and 21% were caused by food consumed in a private home" [7]. Despite this 59% downstream culpability, food processors and manufacturers are generally perceived as the greatest culprit in such outbreaks. This perception may be due to the likelihood of a manufactured product impacting more people across a greater geographic range; it may be because these recalls tend to get the greatest media and social media coverage; or it may simply be because this link in the chain is generally the least transparent, therefore the most suspect (Fig. 9.4).

Regardless of the reason, manufacturers frequently encounter consumer perception of "the buck stops here," making it all the more important that they have a thorough traceability program both up and down the supply chain, and even more critical, that they have verifiable and validated testing, sufficient test and hold, and specified "clean stop" programs by which to limit or disprove culpability in a recall and, when applicable, move the inspection upstream or downstream.

In fact, it is interesting to note that, despite the fact that it is not the most prominent cause of foodborne illness contamination, the manufacturing industry can be said to be the most tightly regulated (although FSMA rules are serving to spread the regulation and responsibility more equitably along more of the food chain).

While time is a significant element of traceability at all levels, its magnitude in the traceability at the manufacturing link may be the most critical aspect of a program. Thus, the manufacture should know its traceability process so innately as to be able to take immediate action if a product is implicated, whether by in-house testing, regulatory notification, consumer complaint – or any other reason.

One of the most common means of achieving this is the practice of holding mock recalls. In fact, while FDA is still in the process of developing FSMA's product tracing rules, the agency has published National Commodity-Specific Food Safety Guidelines, such as that for cantaloupes and netted melons, that recommend that "A trace-back and trace-forward exercise should be conducted at least annually and should achieve accurate traceability within four hours or as required by applicable regulations. The trace exercise should achieve an account of all product one step forward and one step back (100% reconciliation)." [4]

While noting that product tracing systems are not a preventive measure, the guidelines do state that they "are an important element of a comprehensive food safety program and should be verified periodically for effectiveness." With this in

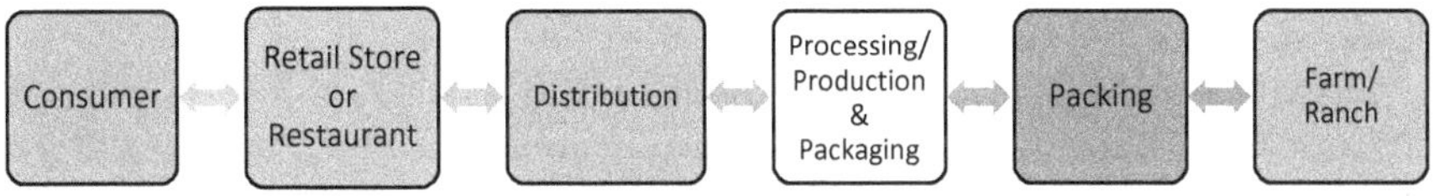

Fig. 9.4 Supply chain: manufacturing

mind, there is a certain likelihood that the FSMA rules on product tracing will incorporate similar recommendations – or mandates.

Although recall exercises are an important element of the food safety program, even more critical is having the alert system to notify the manufacturer of a potential contamination – i.e., product testing – and the retention of product until it is verified as safe – i.e., a test and hold program. This test and hold program should be set up so as to assume that every test will come back positive, and that a process be in place, that every person knows his or her responsibilities if and when it does, and that back-up personnel be assigned if the primary is unavailable. All of which are aspects of ensuring that the critical element of "time" be fulfilled and product be stopped from moving further into the supply chain should a contamination be detected.

Additionally, to halt sales of a product implicated after shipping, the manufacturer must have the ability to communicate with all downstream recipients at all times – from the designated contacts of the national grocery chain to the owner of the mom-and-pop market who may be camping in the wilderness on the Saturday night that the detection occurs, and/or any distributor or broker in between.

Downstream Customers

Such brand protection challenges also extend to the downstream handling and use of one's product. A key, nearly universal, aspect is ensuring that the product is transported and stored at the proper temperature. But there are also commodity and product-specific aspects, such as the beef producer ensuring that bagged beef intended to be sold whole, is not ground at retail – a use for which it hasn't been tested and found safe.

As technology continues to improve, the manufacturer is able to gain greater control over such downstream traceability. For example, there are sensors that can be placed with the product during transportation for remote monitoring. Not only can these monitor the temperature of the product, but with GPS tracking, the trucks can be monitored for location, speed, unscheduled breaks, etc. When drivers know they are being tracked, they may tend to be diligent, resulting in added protection for food safety and food defense as well.

This can be of particular benefit when high-value foods (such as baby food) are being transported, as these products are the most susceptible to theft and diversion. Along with the risk to the quality and integrity of the product, economic adulteration of stolen product takes such forms as product dilution and label recoding, with subsequent sales to small, local, or discount stores that may have less-sophisticated supply standards.

Despite the fact that recoded and/or diverted product is now likely to have been compromised – out of the hands and traceability of the manufacturer – it is still the manufacturer's name and brand that is at risk, along with the consumer. Additionally, it is virtually impossible for the manufacturer to know where the product has gone or what may have been done to it unless and until a customer complains, and it is discovered the product came from a store to which the company never sold its product. At this point, the challenge will be to convince that store owner to trash the product and to only buy from certified/approved brokers in the future.

Upstream Suppliers

General upstream traceability can be just as much of a challenge for the manufacturers, particularly those that produce a great deal of product, a variety of product or multi-ingredient product – which can be said to be applicable of the vast majority of food manufacturers. In such cases, it is likely that lots will be commingled, requiring that every lot number of all ingredients in a product be recorded and tracked.

This is particularly essential when manufacturing specially labeled products, such as organic or kosher-certified, or those not listed as containing an allergen. In such cases the manufacturer must not only follow all correct protocols but must ensure against commingling with non-certified or non-allergen-free ingredients. If a shortage of a certified/allergen-free ingredient occurs, substitutions cannot be made to complete the order. While non-certified substitutions would not impact the *food safety* of organic- or kosher-certified foods, it is a consumer issue that can greatly impact the reputation of the manufacturer – as the consumer is paying for a specific characteristic, and there are consumer groups who will purchase such product off the retail shelf purely for the purpose of verifying its certification.

Similarly, the tracking of foreign ingredients back to the farm can be of significant challenge to the manufacturer, particularly when the ingredients are grown by, and only attainable from, small farmers. The confectionary industry faces such challenge with the purchase of cacao, which, because of these trees' need for specific amounts of warmth, sun, humidity, and shade can only be grown within 20 degrees of the equator, and because of the amount of attention and care required for their growth, are generally grown on small farms of less than 10 acres, and the beans commingled through exchanges before being roasted and processed [2].

While downstream, upstream, and international traceability pose challenges for manufacturers, the integration of their own internal systems can be just as demanding, as it can be difficult to have a single conjoined system that flows throughout all operations and departments of a plant. Because of this, a manufacturer often will have one system at receiving, manufacturing, and distribution – then link these with the others using third-party widespread systems. However, with the variation of data needed for each operation, and differences between plants themselves, third-party systems can still be cumbersome, expensive and difficult to configure to the user's need.

Add to all this the need to continually adapt to ever-increasing regulation, retail and restaurant requirements, global standards, and consumer needs that revolve around trends and perception as much as (or sometimes more than) food safety or quality, and it becomes clear that the most successful manufacturer is likely to be the one with the most accurate crystal ball.

Through Packing/Co-packing: Integrated Separation

Packing, or co-packing, has an array of types and definitions and can be a step in the chain both before manufacturing and at the processing step. From the farm, it can be the sorting and packing of produce by type size and quality; it also can be the minimal process of cutting vegetables into individual portions or party trays (Fig. 9.5).

A packer may be a single brand that has a number of growers with whom it regularly works or individually contracts; it may be a co-packer that packages produce for various brands; or it may simply be the facility that sorts and packs bulk product to move it on to the next step.

As a processing step, as discussed in the previous section, co-packing can be a brand's use of a contracted processor for the manufacture or packaging of some its product. This practice is generally implemented for small or specialized runs, such as seasonal products, focused-marketing packaging, etc.

Thus while it is an essential link in the chain and may be a completely separate link, it is generally such an integral part of the previous or following step that this chapter discusses packing/co-packing as aspects of growing/processing rather than attempting to separate these out.

That said, as addressed in each of the two linked sections of this chapter, there are challenges unique to packing/co-packing, resulting primarily from the communication and integration of processes needed when two separate businesses must, essentially, operate as one.

Packing and repacking of produce can occur at many steps in the chain, from a single brand utilizing multiple growers (addressed in the next section) to co-ops and conglomerates that warehouse and pack multiple products from multiple growers, to repacking at a distribution center to sort out produce that may be going bad or to resort product for a particular use or customer.

While lots will be traceable upon receipt, once the commingling begins, the traceability challenges expand. And when produce is commingled multiple times – e.g., in the packing house, at the distribution center, then at retail to create a salad or fruit platter – the accuracy of the tracking information has multiple opportunities for errors.

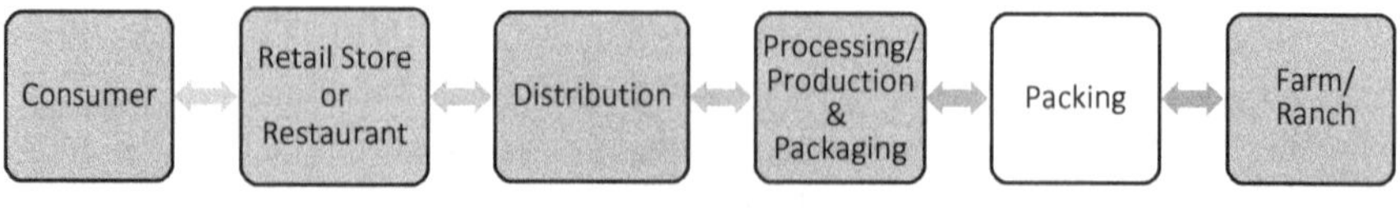

Fig. 9.5 Supply chain: packing

Co-packing can also add to the potential for purposeful economic adulteration. With co-packers identified simply by the code on the package, any alteration of a single number or letter in the code, or the application of a new label with false information, would completely nullify traceability and disassociate the actual packer.

Some of the greatest challenges of traceability in co-packing in manufacturing are that of labeling – ensuring the ability to integrate the co-packer's plant code into that required by the brand being co-produced. In such production, it is the commingling of ingredients that is most likely to introduce challenges. That is, because the runs are generally small, a very small amount of bulk product is likely to be used. Yet even in small amounts, this bulk product, such as rice from a silo, may have been commingled from different lots which can settle together as rice is fed from the silo into production.

From the Farm: Where It All Begins

It all starts at the farm. Thus, a single inaccuracy at this point – whether it be the accidental transposition of a number, the hurried "pencil-whipping" of field notations at the end of the day, or the inability to decipher a hand-written entry – will impact the traceability all the way through the system. This is the same whether it is the tracking of field produce or food animal (Fig. 9.6).

Thus, the greatest challenge at this level can be ensuring the accuracy of information that is passed forward throughout the chain. Although automation is increasing with pallets coded and tagged, systems still require that the user in the field enter the information (e.g., field, day, time, crew, etc.) in one form or another. There is a great deal of potential for the use of GPS devices that can automatically enter much of this information, but the industry, as a whole, is not yet at that level.

Additionally, there are some growers who do still rely on handwritten field tags. In such cases, there is a reliance on the integrity, literacy, and accuracy of the person recording the information; and the need for a dual step of manual entry into a recording system leaves opportunity for further error.

Not only can there be inadvertent mistakes of entry in the field, there can be purposeful mis-entry of information, e.g., if produce is being harvested from a different field than expected, etc. This could have an impact on traceability should a recall be required and be an issue if the produce were then mislabeled as organic or natural.

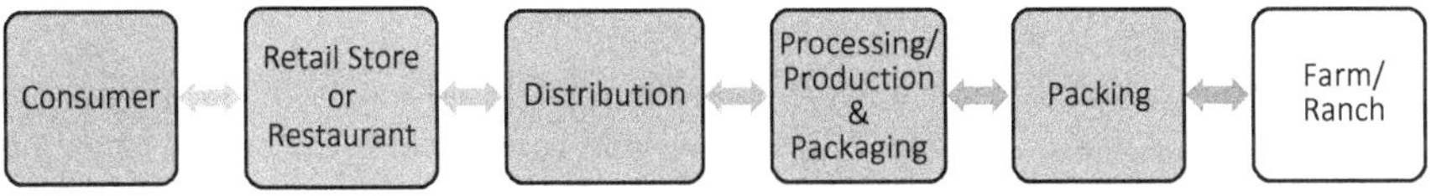

Fig. 9.6 Supply chain: farm

Probably the most organized system of traceability for produce is that of the Produce Traceability Initiative (PTI), launched in 2008, and now sponsored by Canadian Produce Marketing Association, GS1 US, Produce Marketing Association and United Fresh Produce Association, which provides for case-level electronic traceability of produce throughout the supply chain. The system uses a single code to follow the product through the chain all the way to retail and is described in more detail in Chap. 7.

While PTI has been largely adopted by the growing and packing communities and its use is continuing to grow, it has not been fully adopted downstream, primarily because it relies on use of the GS1 barcoding and the ability of the systems at each step of the chain to read and utilize the coding. Additionally, at present, not all distribution centers or retailers have the infrastructure to receive bar-coded information.

As with so many traceability aspects, it is most likely to be the large, resource-rich companies that are implementing and using PTI, while small and local businesses may continue to handwrite their records. This can then become a factor for all packers – large or small, because when produce is harvested and commingled from a number of fields, use of produce from a field that doesn't provide electronic product coding takes that traceability back down to the manual level.

The impacts on the produce industry of commingling of product from multiple growers and its use in complex products is clearly illustrated in numerous recalls. Once a foodborne illness outbreak is declared, and the investigation begins, the first step is determining exactly which ingredient is the culprit. And when a complex product with multiple ingredients is implicated (e.g., salad, salsa, etc.), the specific ingredient has to be determined. And the determination of that ingredient impacts all growers of the item – even if that determination is incorrectly made.

Take, for example, the *Salmonella* outbreak of 2008, in which FDA linked tomatoes to the outbreak and warned the public not to eat certain lots. While the warning was of specific growers and packers, consumers – who tend to be wary of all products/ingredients once implicated – began to avoid all tomatoes, and some retailers and foodservice providers stopped carrying tomatoes altogether. The issue was further exacerbated by the later decision that the source was actually raw peppers from Mexico, not tomatoes at all. Losses were estimated to run as high as hundreds of millions of dollars for tomato growers and packers.

The 2006 spinach recall provides another example of the impact of traceability challenges on an entire industry. Although spinach was correctly identified as the source of the *E. coli* outbreak, eventually leading to implication of Earthbound Farm on which former Chief Food Integrity Officer Will Daniels has frequently spoken, there was still the issue of commingling of the spinach from various fields, so that multiple growers were implicated. At that point it became a subjective call of the investigators as to where to put their focus.

Even prior to that, with the array of spinach uses in finished product, tracing had to be conducted both backward to the source and forward to the potential range of distribution – so the warning went out to consumers to not eat spinach. Period. Not fresh, not frozen, not at all. At one point, it was even thought to potentially be a bioterrorism issue.

Interestingly, Daniels explained that one reason that Earthbound Farm became the primary focus was because the company knew that it could track recipients of its product, so it chose to conduct a voluntary recall. At that point, all attention was turned on it and others dropped out – even though multiple sites had been implicated.

While less commingling of product would have, at least somewhat, reduced the industry impact in both situations, it is not a feasible solution when considering the extent and speed of the market and the current limits of traceability. Additionally, even with the use of a GPS system, traceability remains at the case and/or field level, not the item level. Technology is continuing to improve, however, so that we are likely to see more traceability down to the square block, say 10'×10', rather than to Field X. This would enable tighter investigations with lesser amounts needing to be recalled, but would also require some redefining of a "lot."

Technological evolution is also dependent on the adoption by the food industry and the need for industry to continually challenge and improve its systems. For example, if a grower or packer codes and tests every pallet from a field, rather than considering all produce from a field as one, the definition of lot can be challenged. That is, if only one pallet of four from a field tests positive, the grower/packer can challenge the traditional practice of destroying all product from the field to destroying only that of a single pallet. So that if it were shown – through an accurate testing program of field lots and finished goods – to be simply a sporadic contamination, and a good test and hold program was implemented, the exposure and impact would be a great deal less severe – not only on the grower and packer, but also on everyone down the chain to the retailer and consumer.

That said, food safety as a whole is only as good as the continuum. Even when everything is done right in the field, an error, contamination, or other incident at receiving can cause the loss of the product tracing.

Food Animals

While there are obvious differences in the growing of produce and the breeding of animals for food, the commingling of product creates similar challenges for tracing meat back to the animal as for tracing produce back to the growing field.

Many food animals, such as swine and cattle, are marked with ear tags, to enable coding that follows them through the system. However, once the animal is moved through processing, it becomes more and more difficult to track. This is because once slaughtered, the meat from the animals will be separated by type and grade. The carcasses are generally not scanned at this point, and meat may be commingled from numerous feed yards, so it can be impossible for packaged meat to be linked back to the individual animal.

One significant difference on the farm is that of handling. While this is not so much a factor of traceback for food safety, humane treatment has become a significant factor of traceability for many brands and retailers who are focused on and/or market "ethical sourcing."

Long before ethical sourcing became an advocacy platform, animal handling has been an aspect of concern both for the welfare of the animals and the quality of the meat. As such, transparency on the farm and in processing are becoming ever more essential for consumer communication.

As noted by Temple Grandin, world-renowned animal science professor and livestock handling system designer, such transparency is critical not only to show what the industry is doing right but to counter the misinformation of the "hidden videos" of the internet [6].

It is also needed to take consumers back to the roots of meat production. The beef, pork, and chicken on our tables is – and can only be – produced through the slaughter of animals. In today's urban culture, it all too easy to "forget" that fact and think of meat as originating in its packaged, retail state – with the only ones focused on or publicizing the actual origination of meat being those who are against its use as food, thus wish to show as negative a picture as possible.

For industry's sake, traceability of animal products needs to go beyond that of food safety and communicate, to show and tell the ethical sourcing practices even in slaughter. As Grandin said, "these things need to be more commonplace and ordinary. I think we need to just show it and explain it… we need to show it to the point where it just becomes ordinary" [6].

Summary

When discussing the world's food supply, the word "chain" perfectly illustrates the linkage between the players. Each link of the chain is a unit unto itself with unique stressors and strains, but each has an integral charge to keep the chain intact, each must hold itself interconnected with the next, and a pull on any one link of the chain will impact every other link – forward and back.

Because each link – and each component that makes up the link – is unique, the specific challenges described in this chapter are not necessarily applicable to every farm, packer, processor, distributor, retailer, or even consumer; nor do they depict every traceability nuance of or challenge faced by these separate but integrated links of the food supply chain. But they do provide an overview and some industry perspective on today's traceability through the chain, how the food industry is improving and evolving to increase food safety and meet the needs and expectations of the consumer, and where improvements can and need to be made.

Acknowledgements The following industry leaders lent their expertise to this chapter through interviews and discussions on the challenges and nuances of supply chain link in which their experience lies. The author gratefully acknowledges their contributions and expertise.

David Acheson, Founder of The Acheson Group and former FDA Associate Commissioner of Foods
Will Daniels, President, Produce Division at IEH Laboratories and Consulting Group; former Chief Food Integrity Officer, Earthbound Farm
Bradd Eldridge, Founder and President, Quality Management Partners
Liz Enzor, Director, Food Safety and Health, Global eCommerce, Walmart
Bruce Ferree, Independent Food Safety Consultant; former Compliance Manager, California Natural Products
Brian Honigbaum, Quality Assurance and Food Safety Manager, 7-Eleven; former Director of Regulatory Compliance, Sam Kane Beef
Jorge Hernandez, Chief Food Safety Officer, Wholesome International; former Senior Vice President, Food Safety & Quality, US Foods
Frank Yiannas, FDA Deputy Commissioner, Food Policy & Response, former Vice President, Food Safety, Walmart

References

1. 2018 food and health survey from the International Food Information Council Foundation [Internet]. Washington, DC: International Food Information Council Foundation; c2014 [cited 2016 February 3]. Available from: https://www.foodinsight.org/2018-FHS-Report-FINAL.pdf
2. All in the family [Internet]. Valley View: Quality Assurance & Food Safety (QA) magazine; c2011 [cited 2016 February 3]. Available from: http://www.qualityassurancemag.com/all-family-goetze-candy-quality-assurance-june-2011.aspx
3. Driving the Chain of Custody from Mid-Field [Internet]. Valley View: Quality Assurance & Food Safety (QA) magazine; c2010 [cited 2018 August 7]. Available from: http://www.qualityassurancemag.com/article/-cover-story%2D%2Du-s%2D%2Dfoodservice/
4. National Commodity-Specific Food Safety Guidelines for Cantaloupes and Netted Melons [Internet]. Silver Spring: FDA; c2013 [cited 2016 February 3]. http://www.fda.gov/downloads/Food/GuidanceRegulation/GuidanceDocumentsRegulatoryInformation/ProducePlantProducts/UCM365219.pdf
5. Nearly three-quarters of Americans looking to government for more food safety oversight [Internet]. [Location unknown]: PR Newswire; c2014 [cited 2016 February 3]. Available from: http://www.prnewswire.com/news-releases/nearly-three-quarters-of-americans-looking-to-government-for-more-food-safety-oversight-243648081.html
6. Slaughter plants should broadcast live: Temple Grandin [Internet]. Saskatoon: The Western Producer; c2013 [cited 2016 February 3]. Available from: http://www.producer.com/daily/slaughter-plants-should-broadcast-live-temple-grandin/
7. Surveillance for Foodborne Disease Outbreaks — United States, 2009–2010. Atlanta: CDC; c2013 [cited 2016 February 3]. Available from: http://www.cdc.gov/mmwr/preview/mmwrhtml/mm6203a1.htm?s_cid=mm6203a1_x
8. Walmart announces new commitment to a sustainable food system at global milestone meeting [Internet]. Bentonville: Walmart; c2014 [cited 2016 February 2]. Available from:http://corporate.walmart.com/_news_/news-archive/2014/10/06/walmart-announces-new-commitment-to-a-sustainable-food-system-at-global-milestone-meeting
9. Walmart drives food safety standards [Internet]. Valley View: Quality Assurance & Food Safety (QA) magazine; 2013 [cited 2016 February 3]. Available from: http://www.qualityassurancemag.com/qa0613-walmart-food-safety-standards.aspx

10. Walmart U.S. announces new animal welfare and antibiotics positions [Internet]. Bentonville: Walmart; c2015 [cited 2016 February 3]. Available from:http://corporate.walmart.com/_news_/news-archive/2015/05/22/walmart-us-announces-new-animal-welfare-and-antibiotics-positions
11. Walmart [Internet]. [Location unknown]: Fortune 500 Magazine; c2018 [cited 2018 August 7]. Available from: http://fortune.com/fortune500/walmart/

Chapter 10
Blockchain in Food Traceability

Thomas Burke

Abstract Blockchain is a transformational, paradigm-shifting technology impacting multiple industries. Starting in 2009 with the creation of Bitcoin, the applications of the technology have expanded to a wide range of use cases including food traceability (Abeyratne SA, Monfared RP, Int J Res Eng Technol 5:1–10, 2016). Described briefly, a blockchain is a decentralized, distributed ledger verified through consensus of the network (The Economist, The great chain of being sure about things. 2015). Due to the relative immaturity of the technology, it is difficult to predict how and in what ways it will transform the food sector, but it is clear that Blockchain will be a key technology for improving food traceability systems (Abeyratne SA, Monfared RP, Int J Res Eng Technol 5:1–10, 2016). More broadly, Blockchain is shifting how and what data is shared throughout the food supply chain, moving from siloed, opaque data traditionally held on paper or internal, centrally controlled databases to a more open, transparent system.

Food supply chains have unique challenges which make blockchain data architectures particularly attractive, such as disparate trading partners, hyper globalized supply chains, and unequal adoption of digital technology. The intention of this chapter is to briefly introduce the concept of blockchains and delineate use cases and advantages for food traceability, not to delve into technical computer science. Many traceability-related examples are drawn from current pilots and early implementations of blockchain in the food sector, which include seafood, produce, and meat/poultry.

Keywords Blockchain · Internet of things · Ethereum · Hyperledger · Cryptology

---

Note: For the purposes of this chapter, blockchain is referring to the spectrum of technologies based on Nakamoto's basic premise of a decentralized ledger connected through Merkel trees.

T. Burke (✉)
Global Food Traceability Center, Institute of Food Technologists, Washington, DC, USA
e-mail: tburke@ift.org

J. McEntire, A. W. Kennedy (eds.), *Food Traceability*, Food Microbiology and Food Safety, https://doi.org/10.1007/978-3-030-10902-8_10

Introduction to Blockchain

The history of blockchain starts in the financial technology and e-commerce sectors, an important disclaimer when applying blockchain to other use cases [1, 4]. Originally, blockchain technologies needed no other verification than the network itself, because the assets accounted for by the blockchain only existed on the blockchain [4]. Blockchain technologies were begotten out of an experimental manner of exchanging value, known as cryptocurrencies. The term "blockchain" was first used by Satoshi Nakamoto, a pseudonymous person or entity, in a 2008 paper conceptualizing chronological blocks of data linked through a networked cryptologic chain [12]. The following year, Nakamoto created Bitcoin based on this concept, which is still the most prominent cryptocurrency [4, 12]. Nakamoto's intention of Bitcoin was to create an entity wherein transactions are made without an established intermediary (i.e. banks), which made transactions more transparent and less easily corruptible [12]. Cryptocurrencies have functionalities akin to other currencies but are not tied to nation-states, as fiat currencies like the U.S. Dollar or Euro are [4]. The underlying architecture, blockchain, is able to leverage the capability of a global, open network combined with a cryptologic methodology for generating a secure, trust-less means of exchanging value or information [4, 12].

Since blockchain technology is essentially a database system, it has vast applications across other industries, including product traceability, logistics, and other financial applications [1, 4]. Innovations by subsequent blockchain oriented organizations, such as IBM, the Linux Foundation and the Ethereum Network, quickly developed blockchain platforms with the flexibility to harness blockchain for supply chains [4, 13].

Blockchain technologies are new iterations of an existing concept. Ledgers are an inherent tool of business, and blockchain uses technology to improve on some of the disadvantages of private ledgers, namely by reducing reliance on external institutions in favor of cryptologic proof [4, 12, 13]. In the context of food traceability, blockchain technologies are seen as a formidable tool to enabling whole-chain traceability and transparency rather than the traditional, opaque 1-up, 1-down traceability [17].

With blockchain distributed on a mutually shared network, all stakeholders of a supply chain can be on the same page with traceability information [1]. But more revolutionary is the potential for this information to be available for all segments of the supply chain including end consumers.

What Is Blockchain?

Although the initial iteration of blockchain technologies concentrated on creating non-institutional currency, the technology is essentially a ledger with a wide potential of features, depending on the architecture [1, 4, 13]. For the purpose of this chapter, a *transaction* is any addition or manipulation of information on the blockchain. In food traceability, a food item may undergo an internal transformation and would then be noted on the blockchain. For our purposes, this may still be referred to as a transaction, although no money has exchanged hands.

A blockchain uses hash-based cryptology to assure security and trust [12, 13, 17]. A hash is an encrypted version of a string, or sequence of characters, wherein it is computationally impossible to derive the original without a key [13]. The blockchain has three essential pieces of data: the transaction timestamp, transaction details, and a new hash combining the hash and details of the previous transaction [12, 13]. Each transaction is then distributed throughout the network [12, 13]. Through this process, a continuous encrypted record of the transaction is kept and becomes immutable once added to the blockchain [12, 13].

To verify changes to the blockchain, a resolving algorithm audits the pending transaction after which it is then distributed throughout the network to the shared ledger [4, 12, 13]. Once a transaction has become finalized through this validation process, it becomes a permanent part of the chain [4, 12, 13]. The nodes at which transactions are verified are known as "miners" [4, 12, 13]. Blockchain architectures primarily differ in their choice in resolving algorithms and the degree of openness to miners. Some algorithms prioritize decentralization and anonymity while other prioritize throughput and rapidity [9]. Public blockchains reward miners with tokens, such as Bitcoin or Ether, for performing calculations to resolve new transactions [12].

Users of the blockchain have two keys: private and public. The public key is the means for sending material to a specific individual on the blockchain and publicly verifying their actions. The private key authenticates transactions from the individual holder (Fig. 10.1).

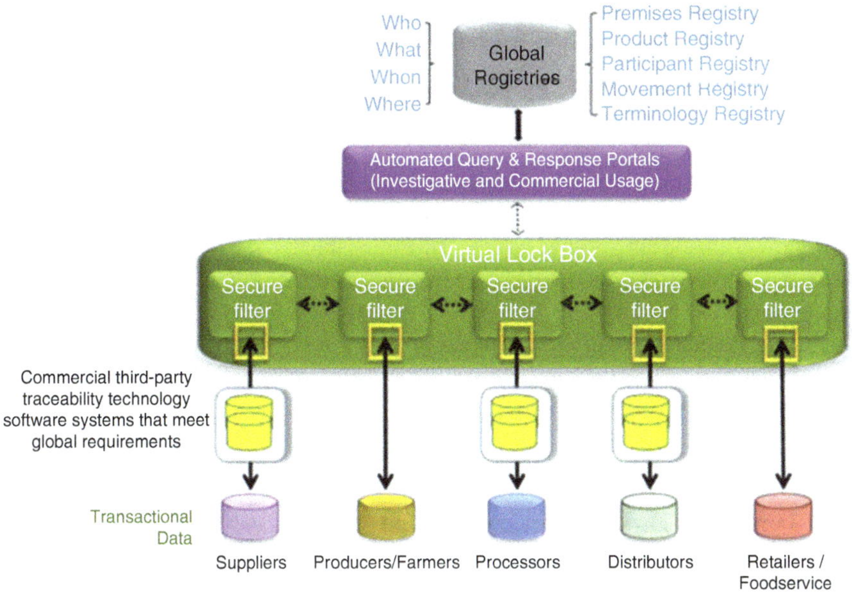

Fig. 10.1 Interoperable architecture in food traceability systems [11]

Although many blockchain applications have been devoted to cryptocurrency exchange, the framework can theoretically be applied to any scenario requiring assured/verified information, including food supply chains [1, 6]. The technology has heavy interest among diverse sectors, for its ability to rely on peer-to-peer networks rather than centralized institutions [6]. By having a more transparent and decentralized system, companies along the supply chain will be able to input data into the system with a degree of anonymity and control that may spur universal adoption [1, 6]. Data verification derived from its cryptologic structure is another attractive quality of blockchain systems.

Blockchains are epitomized by three major components: cryptology, networks, and computation [12]. The mathematics behind blockchains have existed for some time, but widespread high-speed internet connectivity combined with the general increase in computational power have made it possible for blockchain networks to be feasible [1, 12]. By being distributed among peer-to-peer nodes, it is very difficult to usurp the record among all of them [1, 12, 13]. This makes the record immutable, time-stamped, and secure while being trustless in the system ownership [6, 12]. In food supply chains, the distributed nature of blockchains makes it advantageous for food recalls due to the speed with which information is linked.

The value of blockchain use in traceability systems is predicated on the rapidity of querying the system, the simultaneous capabilities of anonymity and transparency, and the immutable and shared nature of the system. The concern with a centralized system for traceability includes a single point of breakdown, the opacity of such a system, and basis on the trust of the provider [1, 6, 19]. Blockchain has the potential for disparate parts of the food supply chain to input data into a shared ledger that reaches both ends of the market, from producer to consumer [1]. Companies can input traceability information while keeping important proprietary or business-competitive information hidden [3].

As of 2018, supply chain and traceability solutions using blockchain technologies have mostly been explored in pilot studies and early implementation [1]. Several companies have started to explore using open-source blockchain bases, such as Ethereum or IBM's Hyperledger, for usage in supply chains [8, 9, 14]. Some of these pilot studies combine other technologies, such as internet-enabled sensors [2, 17].

Many of the benefits touted for blockchain enabled systems are not necessarily exclusive but are rather attributes of strong traceability systems. By using a distributed system that is not implicitly owned by a particular entity, adopting common Key Data Elements [KDEs] may be easier. However, it is possible to have KDEs that are harmonized across an industry while using more piecewise approaches to data collection and dissemination.

Use Cases for Blockchain in Food Traceability

The use cases for blockchain in food traceability are nearly the same as those for general traceability initiatives, which is a primary reason it is so aggressively being pursued by many industry leaders. Food traceability initiatives and technologies are mainly trying to address five primary use cases: food fraud, food safety and recalls, regulatory compliance, social issues, and consumer information. Blockchain in food traceability has the most utility in food commodities (e.g. produce) and disparate, fractured supply chains (e.g. seafood). The utility of blockchains among vertically integrated operations is diminished due to the ability to leverage existing tools in inventory management to accomplish traceability goals.

Food fraud affects all food sectors and has been steadily growing in interest with improved detection methods and greater traceability information being required and available. Michigan State's Food Fraud Initiative describes food fraud activities as "adulteration, misbranding, tampering, overruns or licensee fraud, theft, diversion, simulation, and counterfeiting" [11]. Though food fraud can be unintentional, economic incentives often lead to food and/or information tampering in the supply chain [6, 10, 17]. Blockchain has been seen as a potential tool for combatting the informational side of food fraud [2, 6]. Because blockchain creates a time-stamped, unalterable, distributed record of transformation, transport and depletion, it enables a more straightforward auditing process to investigate food fraud. Previously, obtaining this information would require some compelling reason, such as a food safety outbreak investigation. A blockchain can much more easily be queried and accessed to authenticate transactions or to find the culprits. A particular use case in food fraud relates to seafood, specifically the sale and consumption of Illegal, Unreported, and Unregulated (IUU) fishing [14]. Vulnerable species and ecosystems are being fished to extinction, and over several decades, international agreements have shaped policy on where and when to appropriately fish certain seafood species [14]. However, it is difficult to have a single or interoperable record accompanying seafood as it goes through the supply chain. There are current efforts to use blockchain to resolve these issues, with Provenance being a prominent example [14]. These efforts combine several emerging technologies, like IoT to help solve several issues at once in combatting seafood fraud [6, 17].

Though food traceability has many aspirational use cases, regulatory compliance is the first consideration when devising a food traceability system [17]. The risk of noncompliance can result in unsellable product, fines, and loss of reputation. As regulatory requirements for traceability of food products increase globally, blockchain has a flexibility that would ease and anticipate them [10, 17].

The use case that blockchain most directly addresses are food recalls and safety. As addressed elsewhere in this book, a foodborne outbreak can eviscerate even the largest companies' reputations. Additionally, the commodity killer effect is well known to the industry wherein consumers lose confidence in a particular type of food across the board even against companies and regions that were unaffected by the outbreak. Blockchain provides a decentralized, but unified framework for tracking food as it goes through the supply chain [1, 7, 17]. To support this ideal of whole-chain traceability from source to retailer, a data architecture must be constructed so that there is low cost to each individual supply chain partner, shared responsibility in data stewardship, straightforward interoperability, and security of the record [17]. Blockchain has all of these characteristics, especially public blockchains, such as Ethereum, where transactions can be batch submitted for pennies, reducing a supplier's (whom most often has lowest margins and least amount of resources for purchasing new technology) financial hurdles to enabling traceability. Interoperability between supply chain partners can still be a challenge, especially if they do not agree upon a common platform. However, data import and export out of blockchains are fairly trivial. And most importantly, the auditability and security of the architecture gives them rapid access to the record and can automate much of the often manual process for recalls.

Social issues in food supply chains have existed for millennia, but the power of data access and dissemination have empowered the opportunity to address them. In food supply chains, there are wide-ranging social issues that are as abhorrent as forced labor and slavery to ensuring labor laws are followed to assuring legality of employment [10]. Mainly, food companies are interested in obtaining more information on their ingredients as globalization increases their suppliers and sources [10]. Though blockchain has more limited value in tracking information that doesn't want to be tracked, the advantage of more information is it gives a starting point to investigating social issues. For instance, seafood has vast problems with forced labor in aquaculture farms. Requiring information assuring legal labor (most likely through an external audit) carrying forward on the product through the blockchain would help address this issue [14].

One of the strongest use cases in the startup space on blockchains is increasing end consumer information on food products. The clean label movement and industry data show that consumers are increasingly concerned with the origin, production, and supply chain of the products they consume. Clean labels accomplish this through assuring certain ingredients or additives weren't used, but mostly it is a marketing tactic. However, it does exemplify the consumer's desire to have information on their product. Smart labels are another instance of increased consumer information [16]. Blockchain, by unifying the ends of the supply chain, can give companies the ability to educate their consumers on their product's origin and production [7, 10].

The use cases for blockchain in food traceability are not limited to these instances only, but to expand on all possible use cases would be an exercise in imagination rather than on the current technological landscape. Other possibilities include combining payment and traceability data or NGO certifications (e.g. Marine Stewardship Council, Rainforest Alliance).

Blockchain Configurations

The most visible blockchain environments are known as public blockchains, which are open for any to participate in, provided that they have tokens to post transactions to the blockchain. Most all cryptocurrencies work as public blockchains, such as Bitcoin or Ethereum. However, consensus and private configurations have been implemented which have different properties [18].

There are competing priorities which determine the efficiency and privacy of blockchains [8, 18]. Blockchains have competing priorities based on the use case. To have a truly decentralized blockchain, access is not restricted and transactions are mined based on awarding a token like Bitcoin or Ether. However, as the blockchain grows, transactions take longer and longer to be completed [10]. For supply chains, it may not be advantageous for any person or entity to participate in the network.

For these reasons, consensus and private blockchains have been developed. These architectures retain some of the desirable features of blockchains: immutability, time stamped, and auditability. However, becoming private or semi-private detracts some of the groundbreaking aspects of blockchain and makes it merely another type of database [10]. For many use cases, that is fine, but it increases costs due to having to maintain central nodes, usually at the behest of the service provider [10]. Reduced is that democratization of data responsibility and deinstitutionalization that blockchain promises.

For food supply chains, not having a truly decentralized blockchain network is not critical [6, 10]. Many of the current implementations of blockchain in the food sector are supplied by vertically integrated, large corporations. They have the resources to contract with a service provider to coordinate, convene and host the trusted nodes of the network. These corporations also have the resources to work directly with supply chain partners to ensure best practices and technologies are adopted to effectively carry out the initiative.

Other food supply chain initiatives using blockchain have consensus configurations. So rather than sole ownership of the blockchain being controlled by one entity, the blockchain is shared among supply chain partners. This shares the responsibility of maintaining or paying for transactions to be added to the blockchain.

When considering whether to keep information "on-chain" or "off-chain", the two main concerns are privacy and performance [10]. The architecture of blockchain applications is optimized for assurance of information and decentralization, and thus has a sacrifice when it comes to uploading and transmitting large files. Additionally, in public blockchains, all transactions are visible, so if supply chain partners wish to share business sensitive traceability information, storing information "off-chain" (i.e. in a more traditional, permissioned database) with some linkage on the blockchain may be more advantageous.

Current Blockchain Environments in Food Traceability

Blockchain architectures differ mainly on the way consensus is arrived when adding transactions to the ledger. The two main environments that will be discussed are the Ethereum network and Hyperledger, as those are the two most advanced and useable blockchains for food supply chains.

The Ethereum network is a blockchain environment with wide-ranging potential applications [5]. Though set up as a public blockchain similar to Bitcoin with a token known as Ether or gas, Ethereum can be used to configure networks and even Decentralized Autonomous Organizations [5]. It was created by Vitalik Buterin as an improvement to the Bitcoin concept. He envisioned a blockchain network on which any conceivable application can be created on it. This was one of the first instance of blockchain being used in supply chains.

IBM and the Linux foundation joined together to create a suite of blockchain applications collectively called Hyperledger [9]. There are currently 5 frameworks of Hyperledger, of which sawtooth and fabric are used most frequently with food supply chains. Hyperledger works differently than the Ethereum network, having different resolving algorithms [9]. It uses a lottery-based system rather than proof of work [9]. Therefore, it is better to be used for consensus or private blockchains rather than as a public system.

To Whole Chain Traceability, Transparency, and Beyond

Blockchain is more than a technology, it is also a movement towards greater transparency in commerce [1, 7, 15]. It is important to keep that in mind because the development of blockchain technologies comes from the area of Financial Technology or Fintech and not supply chain, food or agricultural sectors. There are some ideological divides between those developing base blockchain platforms and those whom wish to use it for business operations [15].

Food and agricultural production are among the oldest human activities. Consequently, there are customs and attitudes around the agricultural sector that may not immediately occur to non-food technologists. Agricultural and food production is often dependent on trade secrets: fishing grounds, production methods, etc.

Smart Contracts

Smart contracts are one of the most transformational aspects of a blockchain data storage strategy versus traditional systems such as ERPs [1, 3]. While the idea of smart contracts is not new, with blockchains being tied to value, the value of smart contracts is self-evident. Contracts rely on the exchange of service or goods for currency or some other value [3]. Smart contracts combine the action and motivation for the business relationship.

Smart contracts are programmed to exact financial transactions and business actions to certain conditions [1, 3]. For instance, paying out a purchase order may be able to be executed on the blockchain with minimal human interaction.

Drawbacks and Challenges

The popularity of blockchain applications has revealed some drawbacks that will need to be addressed before being broadly applicable to industries like logistics or food traceability. One is the inherent compromises that exist in blockchain, such as limited transactions per second, which has created bottlenecks in exchanging information on public blockchains [9, 12, 17]. To scale a public blockchain schema for an industry that processes thousands to millions of transactions a second, these types of bottlenecks are unacceptable [17]. There is also an issue with latency, or the time needed to append a block of data to the chain [17].

As with any technology innovation, interoperability will be instrumental in ensuring implementation. For blockchain, that will mean agreeing on a common platform to be used throughout a given supply chain. After all, blockchain is merely enhancing the existing business and transactional relationships in an industry. There will also still be a need for standardizing KDEs.

As with any new technology, there are bound to be speculative businesses using blockchain technology as a dubious value-added service. For an analogy, e-commerce companies proliferated in the 1990s during the dotcom boom, but many made poor business decisions while too heavily relying on the promises of new technology, an infamous example being pets.com. Therefore, if one is investing in a blockchain technology to enhance traceability, it is important to have healthy skepticism on how effective blockchain is being implemented as a supply chain solution. Be wary of any promises that seem extraordinary. Cryptocurrencies are not strictly necessary to using blockchain in supply chains, so be especially skeptical about companies asking to invest in cryptocurrency.

Digitization and Combination with Other Technologies

Blockchains in supply chains are only the data architecture component. To track goods throughout the supply chain, other technologies are often combined with blockchain to accomplish traceability. As has been stated, blockchains in the fintech sector only had to account for assets that only existed on the blockchain, such as Bitcoins. For recording and accounting for goods in the physical world, other technologies have to exist to cover that "first" or "last" mile [1, 2, 17, 19].

Much of the advantages to blockchain stem from the mere digitization of records. There are still many sectors of food production that heavily rely upon paper-based traceability, and in order to have a blockchain system, a company would first have to digitize these records [1]. Data collection technologies, such as embedded sensors or voice capture, are being used in combination with blockchain to accomplish this [1]. There are also robust efforts to use near field communicator (NCF) tags to authenticate and have a physical presence of the blockchain [2].

Conclusions

Blockchain is not a silver bullet solution, especially to the sector of food traceability. Virtually every venture that is using blockchain technologies is still in its infancy, and there are many factors not dictated by technology that are affecting adoption.

However, the potential for improved traceability by way of increased transparency, interoperability, and deinstitutionalization may prove invaluable to finding solutions among the issues in food traceability.

References

1. Abeyratne SA, Monfared RP (2016) Blockchain ready manufacturing supply chain using distributed ledger. Int J Res Eng Technol 5(9):1–10. https://doi.org/10.15623/ijret.2016.0509001
2. Alzahrani N, Bulusu N (2018) Block-supply chain: a new anti-counterfeiting supply chain using NFC and blockchain. CryBlock'18. https://doi.org/10.1145/3211933.3211939
3. Bartoletti M, Pomplanu L (2017) An empirical analysis of smart contracts: platforms, applications, and design patterns. arXiv. https://doi.org/10.1007/978-3-319-70278-0
4. The Economist (2015) The great chain of being sure about things. https://www.economist.com/news/briefing/21677228-technology-behind-Bitcoin-lets-people-who-do-not-know-or-trust-each-other-build-dependable. Accessed 21 July 2018
5. Ethereum (2018). https://solidity.readthedocs.io/en/v0.4.24/. Accessed 27 July 2018
6. Glaser F (2017) Pervasive decentralisation of digital infrastructures: a framework for blockchain enabled system and use case analysis. Proceedings of the 50th Hawaii international conference on system sciences. https://doi.org/10.24251/HICSS.2017.186
7. Francisco K, Swanson D (2018) The supply chain has no clothes: technology adoption of blockchain for supply chain transparency. Logistics 2. https://doi.org/10.3390/logistics2010002
8. Hyperledger Foundation (2017) Seafood in supply chain traceability using blockchain technology. https://www.hyperledger.org/projects/sawtooth/seafood-case-study. Accessed 26 July 2018
9. Hyperledger Foundation (2017) Volume 1: introduction to Hyperledger business blockchain design philosophy and consensus. In: Hyperledger architecture. https://www.hyperledger.org/wp-content/uploads/2017/08/HyperLedger_Arch_WG_Paper_1_Consensus.pdf. Accessed 26 July 2018
10. Lu Q, Xu X (2017) Adaptable blockchain- based systems: a case study for product traceability. IEEE Softw 34:21–27. https://doi.org/10.1109/MS.2017.4121227
11. Michigan State University Food Fraud Institute (2017). http://foodfraud.msu.edu/about/. Accessed 26 July 2018
12. Nakamoto S (2008) A peer-to-peer electronic cash system. In: Bitcoin. http://nakamotoinstitute.org/Bitcoin/. Accessed 26 July 2018
13. Pierro MD (2017) What is the blockchain? Comput Sci Eng 19:92–95. https://doi.org/10.1109/MCSE.2017.3421554
14. Provenance (2016) From shore to plate: tracking tuna on the blockchain. In: Provenance. https://www.provenance.org/tracking-tuna-on-the-blockchain. Accessed 26 July 2018
15. Reijers W, O'Brolcháin F, Haynes P (2016) Governance in blockchain technologies & social contract theories. Ledger 1:134–151. https://doi.org/10.5195/ledger.2016.62
16. Smart Label (2018). http://smartlabel.org. Accessed 28 July 2018
17. Tian F (2017) A supply chain traceability system for food safety based on HACCP, blockchain & internet of things. IEEE. https://doi.org/10.1109/ICSSSM.2017.7996119
18. Xu X, Weber I, Stables M, Zhu L, Bosch J, Bass L, Pautasso C, Rimba P (2017) A taxonomy of blockchain-based systems for architecture design. IEEE. https://doi.org/10.1109/ICSA.2017.33
19. Zheng Z, Xie S, Dai H, Chen X, Wang H (2017) An overview of blockchain technology: architecture, consensus, and future trends. In: 2017 IEEE 6th international congress on big data. https://doi.org/10.1109/BigDataCongress.2017.85

Images

I1. Gooch M, Dent B, Sylvia G, Cusack C (2017) Rollout strategy to implement interoperable traceability in the seafood industry. J Food Sci 82(S1):A45–A57. https://doi.org/10.1111/1750-3841.13744

Chapter 11
Tools and Solutions – Internal Traceability

Judith Kirkness

Abstract Internal traceability is a requirement for food processors, packers, distributors, and others in the food supply chain. It involves tracking all inputs, where those inputs were used in manufacturing (as relevant) and all finished goods made. Further, it requires that records be kept of where all finished products go. This is so you can trace forwards and backwards the history of all inputs and outputs through each facility, in the event of a food safety issue.

While GFSI certification programs require that traceability be in place, they say very little about how to accomplish it. As such, food processors are meeting these requirements with everything from paper systems to sophisticated verified traceability software solutions using a variety of hardware from scanners to tablets to computers and weigh scales with labelling equipment. With so many software and hardware options available, deciding what technology will work best for a business can be a daunting task. This chapter explores the tools available for managing internal traceability and how to evaluate the strength of each solution when comparing options. Using the sample script provided, a comparison of options can be made. With assurance that the one up and one down traceability requirements can be met, software and hardware searches can proceed to compare any additional business benefits the tool can offer beyond recall reporting.

Keywords Internal traceability · Barcode scanning · One up · One down · Software · Hardware · Verification · GS1-128 · EDI · MH10 · ASN

J. Kirkness (✉)
Traceability Matters, Brampton, ON, Canada
e-mail: judith@minotaursoftware.com

J. McEntire, A. W. Kennedy (eds.), *Food Traceability*, Food Microbiology and Food Safety, https://doi.org/10.1007/978-3-030-10902-8_11

Requirements of Internal Traceability Systems

Every food supply chain player is responsible for ensuring that they have "internal traceability" in order to adequately deal with any recalls. To achieve this, food processors must have a record of all arriving raw material receipts including the supplier, item and lot number(s). Whether the processor produces in batches or on a continuous line, they must then track what lots of raw materials go into their interim and finished goods. Finally, the manufacturer must track what lots of finished products are shipped to each of their customers. This may sound like a simple process, however, for many food manufacturers the amount of information can be quite substantial.

There are generally three situations that cause a manufacturer to need to initiate their recall tracing system, whether as a mock exercise or in the event of a real food safety issue:

1. A raw material from one of their suppliers is recalled (or, for a distributor, a recall issued by their supplier)
2. Discovery of a food safety concern with a finished good that reached market
3. Discovery of a food safety concern within the plant, such as a piece of equipment failing inspection

With the required information trail in place, if a supplier recalls a raw material, the company would check if they have any of the recalled lot remaining in stock, and if so, put it on hold for inspection and follow the supplier's disposition instructions. Then they would need to quickly determine what of their interim or finished goods contained any of the affected lot of raw material. Finally, knowing the finished goods lots that may have contained the recalled item, the manufacturer would need to be able to efficiently notify their customers that they are recalling those items and lots from market.

A recall can also be triggered by a problem from your customer or the consumer; in other words, a problem with a finished good that has reached the market. In that situation, the manufacturer needs to both pull that lot from market and launch an investigation to try and determine if the problem extends to other products or batches of that product or other items they produced or packaged the same day or any days the interim product was in the plant, and what caused the issue with the finished goods. It could have been a problem with a raw ingredient, a problem with production, or a problem with transportation and handing by the shipping company, the retailer, or the consumer. To start, if they have any of that lot and item remaining in stock, it should be put it on hold for further testing and investigation, ensuring it doesn't get released to market until it has been reviewed and cleared for sale. Next, there is a need to determine when the product was made and all lots of raw materials that could have been contained in that product. If there is still some of the identical lots of each raw materials remaining, they too can be put on hold for investigation. Finally, the receiving information of what supplier and when those lots of raw materials arrived can be reviewed. If any other raw materials arrived on the same shipment, especially for higher risk items such as those that have to be kept within a specific temperature range, the manufacturer could have those lots held and tested as well.

Finally, investigations and recalls can be launched due to internal findings such as metal detected or a failed swab test of equipment. In those situations, product affected may not have reached market but the investigation may still require forward and backwards tracing.

Most manufacturers will do far more mock recalls than actual recalls, and doing so gives them an opportunity to test their record keeping systems for completeness, accuracy and speed. Even if you can pull together the information in the required time frame, it may still be worthwhile to automate a manual system for two key reasons. The first is to take advantage of the opportunity to use the collected lot information for a variety of business benefits beyond recall. The business benefits that can be derived from improvements in traceability are further described in Chap. 4. The second is that the future of supply chain traceability will require that you make traceability information available electronically, whether by regulation, industry best practice, or to meet consumer expectations, so that an entire supply chain history of recalled items can be assembled from the multiple stakeholders along the supply chain.

The most immediate reason to upgrade internal traceability systems is to take advantage of the opportunity to use the collected lot information for a variety of business benefits beyond recall. How you collect and store lot or serial information in your company will determine what additional business benefits are possible beyond recall reporting. Understanding what benefits are possible will help you determine if investing in new ways to collect and store traceability information provides the return on investment you are looking for.

How Businesses Are Fulfilling the Internal Traceability Requirements

There are four ways that food companies are collecting and storing lot information to meet the internal traceability requirement.

1. Pen and paper logs – the minimum acceptable method
2. Excel or other simple unverified lot recording systems
3. Verified lot/serial number recording systems whose sole function is traceability recall reporting
4. Fully-integrated verified traceability recording systems where the lot information is leveraged for more than just recall reporting

To accomplish this one up / one down traceability and to support internal recall investigations, you need to collect traceability information at various touch points in your process. These touch points are called Critical Tracking Events (CTE). Let's look at each critical tracking event food manufacturers face, and how the above collection systems are used to identify the implications of using that method.

Critical Tracking Events (CTE)

The Receiving Process

Receiving of raw materials is the most important critical tracking event for food manufacturers. This is because, no matter how you track the information, if the information gets recorded incorrectly on the way in, all further tracking within a business will be inaccurate as well. Garbage in, garbage out. It is very hard for any system to recognize if a raw material lot has been entered incorrectly. Generally, most systems will accept whatever lot number you enter or provide to it, as these are numbers generated by a supplier and traceability systems need to be flexible to allow for a variety of lot number configurations. This includes allowing for letters, numbers and varying lengths coming from worldwide suppliers. Making sure that you get the correct lot information from the beginning is imperative. Leading companies hire a smart detail-oriented receiver and train them on the importance of their position in the company's traceability and recall plan.

Key Data Elements (KDEs) are the pieces of information you need to record and keep track of for each critical tracking event, in this case for each received item of food and direct packaging. At a minimum they include the item identification (item code and description), quantity of each item, the lot or serial number for each item, the date the shipment arrived and the supplier the items are being received from. The Global Food Traceability Center has expanded upon the CTE and KDE concepts to further refine how they apply to different commodities and different supply chain points [2].

You can capture these key data elements at receiving in essentially four ways:

1. Pen and paper log
2. Record with pen and paper, then enter into an electronic system
3. Enter directly into an electronic system, e.g. through scanning or tablets
4. Electronically receive the information, e.g. through EDI (Electronic Data Interchange)

Lot Information Recorded With Pen and Paper Logs

The simplest method of tracking the required receiving traceability information is with a paper log. At a minimum, a log at receiving would allow for the recording of the date and supplier delivering and the items, description, quantities and specific lots arriving. If this information is not entered into any kind of electronic system, the logs would need to be filed. During a mock or actual recall, the logs would be pulled and correlated with information kept about manufacturing and shipping to piece together the required history for that item. There is the possibility of transcription errors or issues with legibility using a paper-based system. Depending on the size of your company, increasingly around the world governments are requiring that you hold records electronically and paper and pen systems will no longer be acceptable.

Lot Information Recorded With Pen and Paper, Then Entered into an Electronic System

Many companies record information with pen and paper but then enter that information into some type of electronic system. The capabilities of the system will determine how strong your recall traceability will be and what other business benefits might be achievable from the information entered. This method however still allows for additional errors at the stage of input, since in addition to the opportunities inherent in a paper-based system, transferring the information from paper provides another point where mistakes can be made.

Programs such as Microsoft Excel are easy to use and capable, but spreadsheets are unfortunately prone to errors. Lines can easily be added and moved around. If the cells containing formulas such as totals are not updated, those added or edited lines might not be included. Excel is an example of an unverified traceability system. You can enter any value into a spreadsheet, you can add or change that information at a later date either unintentionally or deliberately. Few are set up to permanently record when an individual cell has been altered.

There are other electronic systems that offer lot number fields you can enter lot numbers into. Some are traceability specific programs and others are part of a larger ERP (Enterprise Resource Planning) companywide software program.

Enter Directly into an Electronic System

Some companies prefer to avoid hand written logs and go right to entering information into an electronic system. This might be by scanning incoming labels or tags, or by using tablets or a workstation at the receiving dock to record information. To understand how scanning can capture the required lot information, you will need to understand the industry standard barcodes that are used in the food industry in North America and many parts of the world.

A Quick Lesson in Barcodes

Any sequence of numbers or letters can be made into a barcode. The value of a barcode lies in what information it can transmit through the scan and what can be done with the information received. The most common barcode people are familiar with is the 1D barcode that consists of a series of vertical black lines on a light or white background. It is called a 1D (one dimensional) barcode because it can only be scanned in one direction. A linear scanner must see all the vertical stripes in a 1D barcode to be able to capture the information contained. Picture the red line of traditional bar code scanners where you need to point the scanner so the red line cuts across all the bars. You wouldn't be able to turn that scanner sideways and read the 1D barcode. There are many formats of 1D barcodes that you will find in the food industry such as GTIN-12 (Global Trade Identification Number 12 digits in length, formerly called UPC (Universal Product Code)) which is the typical barcode on North American grocery items. Not all barcodes are useful for traceability.

As discussed in Chap. 6, in the meat industry, serial numbers are more common than lot numbers. This is because meat processors often sell their meat by what is called catch-weight. That means they price the meat in $/lb or $/kg and invoice their customers for the exact weight of what is shipped multiplied by the price/lb or price/kg. As a result of how they price their products, they put a unique serial number on each and every shipping unit. To do this, the GS1-128 barcode (a barcode format introduced by GS1, the Global Standards Association www.gs1.org) has become the standard barcode in use in the meat industry in North America. As referenced in previous chapters, GS1, a non-profit organization, operates in over 100 countries and has created the most widely accepted standards for organizing and transmitting information worldwide.

A GS1-128 barcode can convey multiple pieces of information that can be captured in a single scan. The company generating the barcode decides what information they wish to include in their GS1-128 barcodes. There are 99 segments, called Application Identifiers, which could be included in a GS1-128 barcode. Typically, these segments are separated with brackets in the human readable information generally shown below the barcode. To decode barcodes on your received items, refer to the GS1 Application Identifier list of the currently used segments found in section 3.2 at http://www.gs1us.org/DesktopModules/Bring2mind/DMX/Download.aspx?Command=Core_Download&EntryId=618.

In the meat industry, the information in the GS1-128 barcode typically includes the GTIN (Global Trade Identification Number or sku—segment 01 or 02), the weight of that item (segment 310 (kg) or 320 (lb)), the serial number for that item (segment 21) as well as a date segment, such as the production date (11) date or the date of expiry (17). Here is an example of a GS1-128 barcode used in the meat industry. It contains four segments—a GTIN of contained items (02), a production date (11), a weight (3102), which is the weight in kg to two decimals, and a serial number (21) (Fig. 11.1).

It's important to realize that with 99 possible Application Identifier segments, just because you see a GS1-128 barcode on an item, doesn't mean it contains the segments that are useful for traceability. Your electronic traceability system will need to have the intelligence to read and separate the segments in a GS1-128 bar-

(02)12345678901234(11)151228(3102)002005(21)1234567890

Fig. 11.1 GS1-128 Barcode used in the meat industry

code for you to use these labels in your company. To be used for traceability purposes, a GS1-128 barcode needs to contain a lot number segment (10) or a serial number segment (21). If it does not contain one of those segments it is still possible to use that label for traceability if your software scanning program has the intelligence to look for and assign as the lot number a date segment such as the production date (11), or the packaging date (13), the best before date (15) or the expiration date (17). Not all electronic traceability systems can use a date if no segment (10) or (21) are provided. It's important to understand the capabilities and limitations of any electronic traceability systems you are considering for your company, so it can handle the type of labels and products you will receive.

Below is an example of a GS1-128 barcode for a food processor which includes just three segments—a GTIN (01), a packaging date (13) and a lot number (10) segment (Fig. 11.2).

Inventory scanning programs have been in use by large companies for many years. However, many of those scanning programs do not read GS1-128 barcodes and cannot be used for traceability. For many years, warehouse management systems focused on just tracking items at the level of the product SKU (stock keeping unit), without regard to a lot or serial number. As a result, the barcode that can be found on many products is called the GTIN-14 (also called ITF-14 or the SCC (shipping container code)). This code embeds the supplier prefix with the item code but does not convey any traceability or date attribute information. Often, companies have this code pre-printed on their cases with the item description and company information. They may then have human readable information on their case from a packaging machine that shows the lot, production and/or expiry or best before date. Below is an example of a GTIN-14 (Fig. 11.3).

Even if a company has a scanning inventory control system, that system may or may not be capable of being used to capture and share lot information, and hence be useful for traceability. While the use of GS1-128 barcodes on trade items is increasing, many food items still do not use this form of barcode. Some verticals within the food industry have moved in recent years to the GS1-128 with segments they deem relevant for their vertical. For example, as part of the Produce Traceability Initiative (PTI), many produce suppliers around the world are now using a GS1-128 barcode containing the (01) and (10) segments. Because the GS1-128 is not yet fully adopted,

(01)12345678901234(13)150328(10)ABC123

Fig. 11.2 GS1-128 Barcode for a food processor

Fig. 11.3 GTIN-14 Barcode

food manufacturers must record and/or enter the lot information somewhere for their internal traceability chain to begin.

While barcodes are still the most common method of identifying items, some companies are using RFID tags or labels instead of or in addition to barcodes. Many new scanners can read both RFID tags and barcodes. Often the choice of what technology to use for scanning is based on a combination of what the company feels is best for its internal tracking, combined with what their suppliers are providing to them and what their customers are requesting and expecting of them.

How to Track Serialized Products When You Don't Have Scanners

Smaller companies who do not have a scanning system but who receive serialized product in their facility, will generally assign a lot number to the entire shipment from that supplier or a lot number to each GTIN or sku arriving from that supplier that day. This is so they can track a single lot number when moving multiple cases around without having to write down long serial numbers for each case. The prevalence of serialization in the meat industry resulted in meat companies becoming early adopters of scanning technology.

Most food manufacturers attempt to retain the use of supplier's lot numbers within their own facility, so that if the supplier recalls an item, the customer can begin looking for the provided lot number and does not have to cross reference it with an internally assigned lot number. Some companies have a policy to always assign an internal number, perhaps so it is shorter or more consistently formatted, so that picking staff make fewer errors. This is commonly seen in distribution, where internally generated "license plate numbers" are assigned often to an arriving skid and are often able to be linked to the supplier-provided lot number or purchase order number. The other occurrence in which manufacturers will assign their own lot numbers is if they are using a scanning system that will only accept numbers and where their suppliers have lot numbers which contain letters. The GS1 standard allows for alphanumeric lot numbers up to 20 characters in length so most traceability systems can accept numbers and letters. If a manufacturer assigns their own lot numbers, it is important that their cross-referencing system be reliable, as the supplier will be communicating the lot number they assigned to that product should they issue a recall.

If you are comparing inventory systems, be sure to understand whether the software is capable of tracking traceability information (using GS1-128 formats) or if the system is only useful for inventory control using GTIN-14 (ITF-14 or SCC codes). Some software systems will allow for traceability but will require separate barcodes for the item code and the lot information if they cannot read and interpret the GS1-128 format. This means that instead of one scan capturing the required information, moving and picking product would require at least two scans, slowing down the warehouse movements.

Receiving Item and Lot Information via EDI (Electronic Data Interchange)

The final method that is being used in industry to make the receiving process easier and more accurate between suppliers and their customers is electronic transfer. EDI, or Electronic Data Interchange, is a method of transmitting information between computer systems using consistently formatted files. EDI works even when the system generating the document is different from the system reading the document. Similar to GS1-128, EDI documents have multiple fields that can be populated and transmitted. So, just because you are using EDI with a supplier does not necessarily mean that lot or serial number information is flowing between the firms. Companies that communicate with EDI are called Partners. This is because EDI can be used to communicate information between any two companies for multiple reasons. It might be for suppliers to place orders with a manufacturer, or for a manufacturer to communicate information with a third-party warehouse or shipping company, or even between two operating divisions or locations of the same company. EDI documents have standard numbers that identify the format. For example, an invoice is an 810, and a PO is an 850.

An ASN (856 – Advanced Shipping Notice, Manifest), is an EDI document. An MH10 shipping label is a specifically formatted serial numbered label that is used in conjunction with the ASN EDI electronic document. Essentially, the shipping partner puts an MH10 serial numbered barcode label on each pallet shipped (called an SSCC – Serial Shipping Container Code). Using EDI functionality with a software program, they record (often by scanning), for each serial number being shipped, what items, lot numbers, expiry dates and any other information their customer has requested about the product is contained on that particular skid. An electronic EDI ASN file is created with the serial numbers and what each corresponds to. It is then typically transmitted through a VAN (Value Added Network—like an electronic mailbox) provider to the partner. Sometimes the EDI file is sent directly to the partner avoiding the need for the VAN. The receiving partner loads the EDI ASN document into their computer system awaiting the physical arrival of the shipment. When the goods arrive, and the partner's scanning system scans the serial number on the MH10 label for that skid, it populates the partner's database with all the details about the products contained on that skid. This ensures accuracy and speeds up receiving time. This method of communicating between partners is common for food companies working with large retailers or large food service providers, as it allows the retailer/customer to receive skids and move them around the warehouse with a single scan of the serialized label, while maintaining the detail of what is contained on each skid in their inventory system.

That same retailer may break the skid and case pick when assembling shipments for its stores. Depending on the sophistication of the retailer's software, they may or may not be able to carry forward the lot information to the individual store they ship it to, especially if you are permitted to ship them a skid with multiple lots on it.

Should Received Items Be Relabeled?

Food manufacturers generally relabel arriving raw materials for one of two reasons. The first is to create a consistent label format and the second is to ensure you have a scannable barcode for an internal software system.

A consistent label format makes it easier for picking and warehouse staff to identify that product without having to look for the item and lot information on the variety of labels that their suppliers are providing. Companies might relabel arriving inputs even if they are not including a barcode or scanning. This is especially helpful if what you call the item internally is different from what your supplier calls it. For example, if you buy the same type of flour from a few different suppliers, you may wish to relabel the incoming product with the generic name FLOUR, regardless of which supplier you buy it from. In each case, your supplier may have their own item code and it can often be easier for picking if staff think of the item in terms of your item codes. This is especially helpful if you are not using scanning to pick raw materials.

If you use a bill of materials or recipe system, you will have a defined item code for each ingredient in your recipe. If that manufacturing system is tied to your inventory system, it may suggest what items and lots to pick using your internal item codes. If you are going to relabel, you could also print the lot number in a large font and position it in the same spot on every label so that picking staff don't have to look for whatever format the supplier has provided that information (such as human readable printed date codes or lot numbers). This could reduce recording errors ensuring the information in your lot tracking system is correct.

You may also choose to relabel all inputs not arriving with a GS1-128 barcode, if you wish to use scanners for picking and lot control within your facility. Relabeling slows down the receiving process but often makes it easier for picking and warehouse staff to handle and track the product after it has been logged in.

Inventory scanning systems that also manage traceability have the information needed to offer a variety of value added business benefits. Buyer beware however--just because a traceability technology has the information needed to offer tangible benefits doesn't mean it has been programmed to use the information for purposes beyond recall. Understanding how the lot information can be used is valuable so you can select the right technology for your business. As we look at the next critical tracking events of moving items around the facility, consuming them in production and shipping them to customers, you'll gain an understanding of what business benefits are possible.

The Storage Process

Moving things around a warehouse is common for a variety of reasons. Sometimes companies will designate a QC or Hold location where product goes until ready to use or ship. Other times, for efficient picking, you may designate default picking locations for items that are replenished as needed. You may have racks with bin locations, or place skids on the floor or on top of one another (canned products have been observed to be stacked two or three high).

The challenge in a warehouse is often where to put things so they can be easily accessed when needed in a way that they are not lost track of. Losing track of products might result in it reaching expiration before being used or reordering when it is not really necessary. At minimum, it causes wasted time and that becomes wasted money. Cycling inventory is important in preventing waste and FIFO (first in, first out) is a common method of managing inventory. Having what you need on hand, but not allowing excessive dollars to be tied up in inventory can be a challenge for many businesses. The added dimensions of expiry dates and lot numbers makes managing food warehouses more complicated.

If you have an inventory system that doesn't capture and report on expiry dates, you may think you have sufficient stock of an item to schedule a product to be made, only to find when you go to pick that raw ingredient, that it is expired and not useable. Not all scanning systems manage expiry dates. If you have fluid expiry dates, such as in the case of fresh fruit, where it could be bad today or tomorrow but requires a judgement call, production planning becomes even more challenging as it could be that looking at the product is the only way to confirm it usability.

Manufacturers will need to decide if there is value in putting barcode slot/bin identification labels to identify each pallet location in a warehouse. Smaller firms who have reliable long-term staff may know where everything is without the need for slot labels. The bigger the company and facility, and the less familiar staff are with where everything is, the more important it is to know exactly where in the warehouse each item is located.

The most common industry standard method of tracking lot information using scanners is with the GS1-128 format. The GS1-128 barcode segments can be encoded into a 1D barcode, a 2D barcode or an RFID (Radio Frequency Identification) tag or label. Many handheld computer scanners can read 1D and 2D barcodes and some can even read RFID tags or labels as well.

Code 128 (GS1-128 is a form of Code 128 barcode) is a recognized symbology that many different software providers have programmed their systems to read and interpret. It can therefore be used for internal (raw materials, interim products) as well as external barcoding (finished goods to be shipped to customers).

Since 1D barcodes are more common in the industry, they are more likely found on raw materials and they are more likely to be helpful to a wider variety of a company's own trade customers. 2D barcodes have a higher successful read rate than 1D barcodes and can carry more information packed into a smaller barcode. RFID is also successfully used in many companies and rewriteable RFID tags offer the benefit that information can be added to the tags as they travel along the supply chain. While you may choose to use 2D or RFID internally, you may want to also put a 1D barcode on your finished goods if your customers only have scanning technology that can recognize 1D barcodes.

Is Live Information Processing Important to Traceability?

Some scanning systems record information within the scanner and transmit that information to the software program when the scanner is docked or in batches when the unit is within range of the base. Other programs are linked live to the system so that information is transmitted as it happens. The advantage of docking systems is that you do not need to have a wireless network established in the areas where the scanners are used, reducing infrastructure costs. However, there are some negatives to batch docking systems as well. It's important to look at what information users need and where they need to get that information to efficiently move product around, pick it for production and prepare orders for shipment.

One important feature for many food companies is ensuring the user is alerted if they incorrectly pick the wrong item. If the user scans an item not on their pick list, many programs can alert the user and not allow them to scan that item. Scanning systems that offer immediate feedback ensure the user does not waste time pulling an item from racking and then have to go put it back when it is discovered that they have the wrong item. Some docking systems will have this functionality; others will not alert the user until the scanner has been docked. If that is after the user has collected and moved the item, substantial time could be wasted in returning the item to correct the mistake.

Similarly, scanning software might be set to look at the expiry date and alert the user if there is an issue. This is possible if the GS1-128 standard is used and the barcodes contain the expiry or best before segment. If knowing that information immediately is important for your business, because you don't want the user to bring that product to the shipping dock or the production area inadvertently, this feature will be important to look for when evaluating systems.

Note however that scanning software can't prevent the user from physically moving or using an item, just from moving or using it as tracked within the system. This is one way that product gets lost in a warehouse, and how the record of what exists in the system can be different from reality. The biggest challenge in implementing scanning technology is typically not the technical part, but the people change management. Staff must buy into the concept that everything they do, such as moving product around the warehouse, needs to be reflected in the electronic system. Staff need to understand that when product is given away as a sample, donated or shipped to replace a damaged product a customer may have received, it too must be scanned and tracked even if it is shipped at no charge. While scanning should reduce errors, it is still important to conduct inventory counts to ensure the inventory counts in your electronic system reflect reality.

Offsite Storage

Companies often use offsite storage as they grow. Perhaps they store raw materials until needed or store finished goods, either returning them to their main location for shipping or shipping directly to customers from the storage facility. Either way, most of those companies want to keep track of what goods they have in their offsite storage locations. In the food industry, it is equally important to know the expiry dates and lot numbers for the items you have in various locations. This is to avoid products expiring in storage before they are used or sold. Also, some customers may have a narrower range of acceptability of product than others. For example, Wal-Mart often asks its suppliers to ship them product that has 75% or more of available shelf life. This request has meant that FIFO is no longer sufficient for inventory control for food companies. If the oldest lot is always suggested for picking, a system that can't check the expiry date against criteria may suggest that you ship Wal-Mart product that ultimately won't match their requirements, resulting in fines or a declined shipment. For these reasons, choosing a software system that allows for multiple location management, expiry and/or lot tracking and allows for customer acceptance criteria, not just FIFO lot suggestion, can be important.

The use of GS1-128 barcodes and potentially MH10 label/ASN EDI document combinations can make communicating with a warehouse easier. Suppliers are often expected to know what customer ultimately receives what lot of product, even if a third party shipper or storage facility performs the actual delivery (the facility shipping the product is responsible in the eyes of the FDA as discussed in Chap. 2, but businesses generally communicate "brand to brand" rather than "facility to facility"). As such, communication with shippers or storage facilities to ensure accuracy of lot information will be important to reducing the scope of any potential recall.

Since banks often want to see regular financial statements from companies to whom they offer operating lines of credit, part of that request will be an inventory valuation. This is possible if your inventory management system is tied to your purchasing and receiving system. Many ERP systems offer this integration, meaning that determining a cost for received items is possible. ERP systems will often have capabilities beyond traceability, allowing for landed cost calculations to be possible. A landed cost incorporates more than just the unit cost paid to a supplier. It includes additional costs to get that product to your dock, such as freight, and duty, brokerage and foreign exchange for items purchased outside the country. Knowing costs for received and manufactured items is critical to monitoring profitability across items and customers. Look beyond basic traceability when comparing systems to see if costing is available and what is included in the costing for that system.

The additional tracking of lot information makes costing even more accurate. For example, if you receive butter at $x/kg this week and $x + $.05/kg next week and you have tracked what amounts of each lot of that sku you have used versus what is still in inventory, a computer can quickly compute a value for that inventory on a live basis. If you have to write off some inventory because it needed to be discarded, knowing exactly how much is being written off based on the cost of that lot when it arrived is possible. Cost calculations and inventory valuation based on lots is a potential business benefit of a fully integrated traceability and ERP system.

The Manufacturing Process

The next critical tracking event occurs when raw materials are depleted and make their way into manufactured products. When the raw materials were received into stock, the traceability chain began.

Distributors have the luxury of just tracking lots in and lots out. While this should be straightforward, some companies employ processes that introduce complexity.

But it is the added dimension of manufacturing, whether it is repacking, assembling products or disassembling product that truly creates complexity in the life cycle of an item. Managing lots through manufacturing is often the step that many traceability systems don't do well.

Similar to double entry accounting systems, verified traceability systems start with the receiving information and require that every transaction involving those lot-controlled ingredients makes sense from that point forward.

If you had 50 units of a given lot in stock on day 1 and tell the system you are picking 30 for a particular production run, it should leave 20 in stock. If you try to then make the same recipe the next day and try to remove another 30 units of that lot, you want your system to alert you that you only have 20 units of that lot in stock forcing you to enter or scan another lot to make up the 30 units and not allow you to use more than you had received overall.

Scanning systems that can read GS1-128 barcodes with lot information can help, as users won't be keying the information; instead it will be gathered through the scan. One of the disadvantages to 1D barcodes is that they have a high non-read incidence. If a 1D barcode is box cut or was applied so it is wrinkled, it may not be readable by the scanner. Users will be faced with keying a complex and very long sequence of numbers into the scanner, which is time consuming and provides a potential for errors. It is important that traceability systems prevent a user from doing something that couldn't really happen. This necessitates getting information into the system in the order in which it really happens. For example, in a verified system you must receive a raw ingredient before you can consume it in production and you must produce an item before it can be shipped it to a customer. It is obvious that this is the real sequence in which things occur in the plant but sometimes due to lack of available people, even with a system, getting the information in the system in a timely manner can be a challenge.

It is only at the transaction point that system checks and balances are useful. If you are writing down lots before entering them into a system, you could use a raw ingredient before it shows in the system as received. If you need to record production so you can assign a finished good lot and ship a product out and your system allows you to use lots that are not already in inventory, it is not a verified system. For good traceability, information needs to get into the chosen system in the order it really happens. That is why scanners, tablets and integrated computers and equipment are showing up more and more on production floors. They ensure timely entering of data as it is taking place, instead of trying to decipher different handwritings later on.

Assembly manufacturing typically takes place in one of two ways. The first is "batch", where raw materials are picked specifically to make a batch of product. The second is "continuous", where raw materials are stocked at the end of a continuous production line and used as needed for the different items being made that day. A good example of a continuous line is a bakery that makes a variety of cake and cupcake sponge mixes, some white, some chocolate, some lemon, etc. They all contain some common ingredients such as baking soda, flour and baking powder, but they also contain some ingredients specific to each item, such as chocolate for the chocolate mix and lemon flavoring for the lemon mix. Some companies will have a portioning process where raw materials are weighed, measured and packaged for easy recipe assembly. This may be referred to as "pre-batching". Others will keep a stock of the common ingredients in the mixing area and clean the holding containers between lots of product. They will deplete these common ingredients essentially by FIFO until they change lots.

Knowing what lots of ingredients are used in batches of finished products is valuable for more than just traceability. One of the key benefits available from lot information is the calculation of costs for interim and finished goods. If the system knows the quantity of each lot used in a batch and had a good landed cost stored for that lot of product/ingredient, producing a cost for the batch is an easy thing for a software program to do. Not all traceability systems use the lot information to calculate costs, so if knowing how much your items cost to manufacture is important for you, be sure to check for that when evaluating systems.

Disassembly manufacturing is what some meat, fruit and vegetable manufacturers take part in and traceability is often a challenge in these environments. For example, a chicken processor breaks down an individual chicken into component pieces and in some case produces waste, such as chicken bones, in the process. Essentially one input yields multiple outputs. The types of products that result from fresh chicken may vary over time, including from day to day, depending on customer orders. They might have several orders for boneless skinless chicken breasts one week and bone in, skin on chicken breasts the next. Fruit and vegetable processors take in fresh produce and may pit items like cherries or wash and destem items like tomatoes. Again, they have one input ingredient and various outputs, some of which are waste. Traditional manufacturing systems were not made for managing disassembly processes well. Some try to accommodate this need using negative numbers; however, negative numbers and lot control don't mix well as the following examples will show.

Traceability systems for disassembly processes allow for multiple outputs from multiple inputs (MOMI). For example, you might input a chicken and some labor and get out two breasts, two wings, etc. For traceability you have an incoming lot of product and need to maintain a connection with all the outgoing products created. How tight this connection gets is often limited by what is practical in your manufacturing environment. In a chicken deboning and manual chicken processing operation, skilled workers cut and debone chicken at an amazing speed. With margins in the food industry so tight, speed is important and slowing down the process for tracking can affect the viability of the business. In this situation there may be many workers at a series of cutting tables which are not washed down between chickens, nor would that ever be practical. So, for the purpose of traceability, any chicken that got cut up on that table could have been exposed to any chicken part previously on that table. Therefore, it is really only necessary to track all the inputs to this cutting process as well as all the outputs from it and to tie that information together for traceability recall reporting. It means that if some problem was found with any chicken item from that day and area, all the chicken items from that day and area would need to be recalled. There is little value in tracking to the individual chicken level in this example.

While this broadens your recall, it is typically what would be required by inspectors anyway; meaning, inspectors would ask you to recall everything that could have been exposed to the problem lot between cleanings. Tracking any deeper would be for internal reasons, perhaps to compare the output of two workers, or for external marketing, if you tell your customers that you can identify exactly what farm or what animal their meat comes from. For most food manufacturers, the speed of production would make that impractical.

Costing of disassembly items is more challenging than assembly but can be a real business benefit of traceability systems that include costing functionality. In assembly production, systems just add up the cost multiplied by the quantity consumed for every item used in the recipe and divide by the quantity produced to get a unit cost. For disassembly, there will be a total input cost that has to be allocated among the multiple outputs produced. There are a variety of ways that systems can compute these costs, such as allocating the input cost in proportion to the relative sell prices of the output goods. The important thing from a system perspective is that costs don't disappear. The input cost and the output cost must match.

Another opportunity for business benefits beyond recall is related to Quality Control (QC). Your QC staff may perform tests on samples of raw ingredients, interim or finished goods throughout the manufacturing process. The results of these tests pertain to the specific lot of product tested. If your traceability system allows for the recording of QC information about a lot and ties that information to the lot as it travels through your facility, it can make producing certificates of analysis easy. A certificate of analysis (CofA) is sometimes required to be sent with your products to your customers. This is especially true for ingredient manufacturers whose finished products are the inputs for other manufacturers further down the line. Only once it is known what lots are being shipped to that customer is it possible to identify the correct CofA to send. This could be an automatic part of the shipping process if the test result information is already tied to each lot of product.

How Should You Label Your Finished Goods?

The information on a finished goods labels will be affected by requirements of the purchaser. If you sell to restaurants or food service, you may or may not include a nutritional panel on the outside of the cases or shipping units. Customers may request that shipping units are marked in a particular way to help them more easily receive and manage product. The GS1-128 barcode is the most common method of communicating information along the food supply chain and as long as a (10) or (21) segment is included, customers with traceability scanning technology will generally be able to read and use those barcodes. The other segments that should be included are the (01) GTIN or (02) GTIN of Contained Trade Items and a date segment such as date of production (11) and/or expiry date (17). There may be other relevant segments to include such as a weight segment (31 – kg) or (32 – lb) and potentially Country of Origin (422). For exported product, Country of Origin is becoming more popular as particular diseases from animals and/or produce are sometimes found in specific countries and knowing the origin of those items can be helpful in providing customers peace of mind.

The Multi-day Manufacturing Challenge

No matter whether you produce in batches or on a continuous line, all finished goods need to be assigned a lot or serial number. Many companies base the lot number on a Julian date, a production date or an expiry date. While some food manufacturers start and finish making their finished products on the same day, many others take multiple days to finish their products. Some products such as beef or sausages require curing for several days, whereas for aged cheese, it can take years for the product to be ready for market. Traceability over multiple days (or years) is another area where some systems fail.

It is important that interim goods in process at the end of each production day be given a lot number that becomes an input for a future days' production. This might involve weighing an interim good that is in bulk form or counting subassemblies to record quantities from those initial stages of production. If production is not complete within one day, be sure to choose a system that allows for multi-level manufacturing with multi-day traceability.

When tracing backward from a finished good, the system should report all products finished on the problem date plus any subsequent manufacturing processes that were not completed until a later date but that were started or used in some way on the day in question.

The Shipping Process

The final tracking event for a manufacturer in the one up/one down traceability process is tracking what lots of finished goods went to each customer.

Some companies fulfill this requirement by indicating on a pick list which lots were picked to fill the order. Since customers increasingly want to receive a list of lots received with their delivery, this information may be entered onto invoices or more commonly bills of lading that will accompany the delivery; it may also be compiled in Word or Excel or in some disconnected system. If the system is manual, you would need to retain the paperwork that contains this information.

The picking process may also be assisted by an inventory control system that manages lots and/or expiry dates. The system might have the capability to direct the picker to the location in the warehouse containing the oldest lot of product, which assists with inventory rotation. This FIFO lot suggestion may be sufficient for most customers, but as mentioned earlier in this chapter, some customers might have a narrower range of acceptability. For those customers, FIFO may not be sufficient to ensure you ship only product that will satisfy their criteria (like Wal-Mart asking for product within 75% of shelf life).

Traceability can also be used in tracking lent assets. For example, a juice concentrate supplier to bars may lend assets to customers as long as they buy their supplies from that supplier. The concentrate supplier may wish to serialize each dispenser, in order to track the asset and where each unit is, even if they don't charge the customer for it.

Getting the lot information into a traceability system might be done by keying it or through the use of scanners while picking. If scanning, traceability systems may offer additional error prevention. An intelligent inventory traceability system can check for any problems live with the pick. Naturally, not all systems will assist with preventing all errors so it's good to understand what error prevention is possible and to look for a system to help prevent the most common errors. For example, a system might prevent the user from picking the wrong item by restricting successful scans to only those items on the order. The system might also check the expiry date or the production date for satisfying customer acceptance criteria or to ensure the product is not expired.

Who Is Between You And Your Customer?

One consideration for recall reporting is whether there is any third party in between the manufacturer and their customer. Food manufacturers often use distributors who may perform repack services for a variety of reasons. One such reason might be to sort through fresh fruit, separating rotting fruit from the fruit that is still sellable. The sorting company might take a skid of 50 cases and pare it down to 45 good cases with the rest going to waste. Another such distributor service is the repacking of combo product (like often found at warehouse stores) or the creation of Point of Purchase (POP) displays for retailers. POP displays can often be found at the ends of aisles and contain mixed items. For example, a half skid POP display might offer a special price on jam, where different flavours of jam are put together into an appealing consumer display. This often involves breaking down complete skids of product, removing them from their cases and combining them in a new way. These processes may be done by temporary warehouse workers and might not involve tracking which lots of product were placed on which assembled display.

Remember that it is often the responsibility of the manufacturer to know what customers received what lots of products, even if a third party is assisting with repacking or distribution. As noted in Chap. 2, the traceability regulation stemming from the Bioterrorism Act of 2002 does not require distributors to maintain records of lot numbers. If your distributors send product to your customers on your behalf, and they provide some value added services to you, you'll need to work with them to devise a system to notify you of what lots of product ultimately get shipped to which of your customers for your traceability record keeping.

If you ship to large food service providers or to the warehouses of many large retailers, you may be asked to provide item and traceability information to them electronically. This will generally be in the form of an EDI Advanced Shipping Notice file (ASN) combined with an MH10 serialized shipping label applied to each skid. Generating the ASN and the matching MH10 labels and getting them on the right skids can be accomplished through scanning or through a more manual process using an online EDI program. This process should be completed in, or as close to, the shipping area, as possible and ideally be done by the shipping staff. One of the disadvantages of paperwork being completed by one person that ultimately gets entered into a system by another person is that dual responsibility often doesn't work well. When one person is responsible and will be called out for errors, they are often more careful as they are held solely accountable.

Whatever traceability system is used should have the restriction that it will only allow for shipment of lots of product that have been recorded as having been made. Basically, the system shouldn't allow shipment of something that doesn't exist or where there is not enough quantity of that lot in that location to make the transaction valid. This ensures that lots of inventory are appropriately depleted and also provides an extra step for error checking. If using a live scanning system, lots may be able to be placed on hold, such as during a QC investigation or until finished good testing is complete. Even if you don't have the space to separate that product in your warehouse, if a lot is on electronic hold, the user could be prevented from shipping that product in the system. The faster a user determines something is on hold, or that they have the wrong item, the less chance of that product being pulled from the shelf, being staged and ultimately leaving the facility before it was ready to.

Clean Traceability Data

Up until recently, a lot of companies had theoretical traceability, but not practical traceability. Theoretical traceability means that the company is capturing lot numbers at each important stage in their business process, and they have a system that contains all this information, but that the data are not actually useable. Practical traceability means that they not only are capturing the data, but that the data exist in a useable form. In many cases, decision makers believe they have practical traceability, because their staff tells them so, and appear to be performing the right steps. But the data are actually unusable, which they may only find out when it is too late – when they have a recall or customer audit. In order to have practical traceability, the system gathering the information must process it in certain key ways.

In theory, the function of a traceability system is simple: users record the lot number, item number and quantity of each inventory transaction, and the system matches them up to flawlessly track the movement of product through the facility. The user can easily see what is presently in stock by lot, as well as see the entire history of a lot, as it moves through the process. But the reality can be quite different. Some critical factors must be present in order for the system to actually function that way.

Editing Traceability Transactions

The first key requirement is that a system support editing of transactions, even after they are completed, and the product has been used. Let's consider a simpler buy/sell example to understand why this is important. Consider a situation where a receiver records that he received 50 of a given lot and then 20 of that lot is shipped to a customer, but then it is discovered that the receiving should actually have been for 48. This is not an uncommon business situation: the supplier could have put fewer on a pallet than normal, the received could have miscounted, etc. With a system that supports editing, the user can change the original quantity from 50 to 48. But many systems do not support editing and the user must instead enter a correcting entry. In this case, for −2 units. But correcting entries are harmful to traceability, as the history for that item will now show:

Lot	Quantity	Date	
History for item ABC-123			
L001	50	May 1, 2015	Receiving
L001	−2	May 1, 2015	Receiving
L001	−20	May 2, 2015	Invoice

Where it should show

History for item ABC-123			
L001	48	May 1, 2015	Receiving
L001	−20	May 2, 2015	Invoice

That seems harmless so far. But it is not, because mistakes can be made with the correction that cannot be made with an edit. Suppose on the correcting entry, which is really just another receiving with a negative number, the user enters the lot number with the letter O, instead of the digit 0. Receiving is the one spot where systems often let you enter any lot number to allow for the wide variety of alphanumeric lot numbers any of your suppliers might provide.

Then the history would now show this:

History for item ABC-123			
L001	50	May 1, 2015	Receiving
LOO1	−2	May 1, 2015	Receiving
L001	−20	May 2, 2015	Invoice

Again, this seems harmless. But it is not, because when it comes time to ship the product, assuming your system is checking to ensure that you can only ship lots that you have in fact received, the user must correctly enter the lot numbers shipped. In this case, the same lot of product is entered in two different ways, so when the shipper records that they shipped the remaining 28 units, the history will look like this:

History for item ABC-123			
L001	50	May 1, 2015	Receiving
LOO1	−2	May 1, 2015	Receiving
L001	−20	May 2, 2015	Invoice
L001	−28	May 3, 2015	Invoice

But where the problem becomes evident is the list of what is in stock. In this case, nothing is in stock, it has all been shipped out. But when the system adds up the stock by lot, this is what is left.

Stock of ABC-123	
LOO1	−2
L001	2

It is obvious that there is no LOO1, because it shows as negative. But it is not so obvious that there is no L001, particularly if there are a large number of lots in this situation. So now the lots-in-stock information is not correct, and cannot be trusted. Not being able to trust your inventory numbers has large consequences for purchasing and for scheduling production, not just traceability. The purchaser might order more when it is not necessary and the scheduler may schedule production when they think they have raw materials that don't actually exist.

Controlling the Lot Inventory

It is very important that traceability systems control the lot inventory, and only allow transactions that are physically possible. Consider our example above. Suppose the lot was received as L001, but then the shipper tried to ship it, recording the lot as LOO1. This is not possible: you cannot ship lot LOO1, because there is not any of that lot in stock. But if the shipper makes the mistake, and the system allows it, the data will be corrupted. The lots-in-stock report would show this:

Stock of ABC-123	
LOO1	−50
L001	50

When actually you have none; and suppose there was subsequently a recall on lot L001, the history would show this:

History for item ABC-123			
L001	50	May 1, 2015	Receiving

But this is the big problem: now we know there is a problem with the lot, and we have no idea where it went. The system thinks it is in stock, but we can physically see that it is not, so we will fail our recall.

It is very important that your traceability system verify each transaction, and make sure that it is all logically consistent and this becomes very complex when the system supports editing, as it must. In this example, the system should have refused the shipment of lot LOO1, because that lot does not exist. The system should have allowed only shipment of 50 units of L001, it should have prevented shipping of 51, or shipping of any lot not in stock, or shipping of this lot on April 30, before it was received. Without the proper control, the data can become corrupted very quickly, and useless for recalls.

Can The Use of GS1 Barcodes And Scanning Ensure Clean Traceability Data?

Scanning barcodes can certainly reduce the incidence of errors, but it does not eliminate them. Simply by failing to record a physical movement or recording a movement without physically moving the product will mean that the location of lots as they appear in your system will be incorrect. In each case, if the system cannot detect that it is logically inconsistent, the data will become corrupted and useless for good inventory management and recalls.

But won't the data be ok, as long as most of the transactions are recorded properly? How harmful could a few errors be? The answer is very harmful! Everything must be done perfectly, so that all the movements in and out match up. Any time the control on a lot is lost, some of that lot is now left in the system, and some is unaccounted for. Each of these orphan lots stays in the data forever, unless you manually clean it all up, and it is just a matter of time before the bad data dominate the history for that item.

How can you tell whether your system is performing properly in this area? One good way to start is simply to look at a lots-in-stock report for your lot controlled items. If there are lots listed there that do not exist, or even worse, negative lots, that is an indication that there could be a problem. It may not be a system problem, it could be that the users are not operating the system properly. Either way it is a problem that must be solved in order to attain proper traceability.

Appendix 11.1 offers a traceability system testing script that can be used to evaluate traceability systems and to determine if the system is going to capture the required information and validate the data at each step after receiving.

Benefits Of Using Electronic Traceability Information For More Than Just Recall Reporting

This chapter has outlined some of the additional benefits that could come from using lot information for other purposes. Below is a summary of the key areas where this information could provide additional value [1], as is elaborated upon in Chap. 4:

- Inventory control – reduce waste, improve rotation of inventory, and order optimally by trusting your counts
- Production planning – using expiry dates to determine what you have available to use

- QC information recording
- Cost calculations
- Tracking assets loaned to customers
- Monitoring yields – with production data over time you can monitor waste and measure performance
- Ensure customers receive the product that they want fulfilling their acceptance criteria
- Profitability management reporting
- Allow for electronic sharing of information – such as with EDI ASNs
- Improved business planning and decision making from better quality information

How To Compare Traceability System Options?

How the lot information is collected, verified and used within the traceability system will determine what business benefits are possible for your business beyond just recall reporting. It's important your internal traceability system also be able to make available the information for outside systems to ultimately tap into. As the industry moves toward full supply chain traceability your internal system may have to interact directly with outside systems, either in the case of a food safety investigation or as part of allowing you to deal with and sell to larger companies. Some industries such as primary meat processors already provide data to centralized government databases. I expect the trend to outside consolidation to continue. The Traceability Matters Assessment Tool was created to help rate traceability technology options for those extra benefits [1]. This Excel spreadsheet contains a series of questions to help rate existing and potential traceability solutions. It is scored out of 100 and through the questions one can determine how well the information will be controlled in that solution as well as how many business benefits the system can offer beyond recall reporting. Also included is an ROI calculator for helping evaluate a payback for investing in new traceability technology. You can request a copy of the assessment tool at www.traceabilitymatters.com.

Appendix 11.1

Traceability Testing Script

Testing a potential traceability system to identify the level of internal controls for maintaining your data is an important step in selecting the right system for your business. Below is a sample test script that can help you identify possible areas of weakness in systems being reviewed, so you can be sure to get the functionality that will be important for you.

Traceability starts at the receiving of raw materials, continues with the movement of those goods around the warehouse and the picking of them for production. The manufacturing process transforms the raw materials into interim and finished goods. The final step in the internal traceability process is tracking what lots of finished goods go to which customers.

Below is some sample data and a generic script to help assess traceability software abilities and identify gaps so that you can better compare different offerings.

Preliminary Set Up

1. **Set up item masters for the raw material ingredients that go into CHEESE:**
   I. SALT→ OE unit: bag; stocking unit: kg; 10 kg/bag; SUP01 $1/kg; if system has scanning, GTIN-14 is 01234567890128
   II. CULTURE→ OE unit: bag; stocking unit: kg; 10 kg/bag; SUP01 $1/kg; if scanning GTIN-14 is 45678901234560
   III. MILK→ OE unit: liter; stocking unit: liter; FARMER $0.50/liter, assume no barcode is available at receiving
   IV. LABOR→ $12/hour
2. **Set up a Supplier called SUP01, whom you will buy the raw materials from. If your system does costing, establish the costs for SALT as $1/kg, CULTURE as $1/kg, and MILK as $0.50/liter.**
3. **Set up an Item Master for an interim/finished good, CHEESE. Each package of cheese is approx. 10 kg, but it is priced by the kg. The list price for this item is $5/kg.**
4. **Set up a Bill of Materials (recipe) for making CHEESE. It is stocked by the package and contains:**
   1 package of CHEESE should include
   I. 100 L of MILK
   II. 3 kg of CULTURE
   III. 0.2 kg of SALT
   IV. 1 unit of LABOR (if the system allows overhead to be included)
5. **Customers like STORE order CHEESE by the case, the sell unit CHEESECS contains two packages (20 kg) of CHEESE per case and is priced by kg.**
   - Set up a Bill of Materials to package CHEESECS which contains:
     - 2 packages of CHEESE
     - Labor – 5 min of the packager's time

6. **Set up a customer master for the grocery store client and give it the customer code STORE. This customer receives a special price of $4/kg for item CHEESECS.**
7. **Set up two warehouses in the system, one called MAIN and the other called COOLER**
8. **Set up two skid locations within the MAIN warehouse for slot A1 and slot B1**
9. **Set your system date to June 3, 2015 (so that the dates used in the exercises make sense)**

Orders, Requirements and Production Scheduling

1. **Enter a sales order for a manufactured item***
    - Customer STORE places an order for 5 cases (100 kg) of CHEESECS.
    - CHEESECS is ordered by the case but priced by the $/kg.
    - Customer STORE has a customer special price of $4.00/kg.
        - Did the system alert you that you don't have any CHEESECS in stock?
        - Does the system produce a pick list for this order, even though we haven't got any in stock yet? If so, it won't be able to suggest lots based on FIFO right now as there is not any inventory available. Ideally, any pick list should have the option of being generated on the day of or day before (closer to when) you make the product and have the raw materials in inventory.

*If your system does not allow you to enter a sales order, how does it tie the lots shipped to the customer?

2. **Review Requirements –what do you need to make**
    - Does the system have requirements reports to help you see what you need to make to fulfill your customer orders? If so, does that report now show that you need to make 5 cases of CHEESECS so that you can ship customer STORE their order?
3. **Schedule Production – schedule to make 5 cases (100 kg) of CHEESECS to fill the above order (Does the system have a date option so you can schedule production into the future?)**
    - Once you schedule to make CHEESECS, review any available shortages report to find out which raw materials you don't have enough of to complete that planned production.
    - You should need to order 1000 L of MILK, 3 bags of CULTURE and 1 bag of SALT to be able to complete this planned production.
4. **Enter a purchase order (PO) to your supplier(s) for any raw material ingredient(s) that are needed to make the finished good. ***

- Enter a PO to supplier SUP01, for 3 bags of CULTURE and 1 bag of SALT.
- Enter another PO to supplier FARMER for 1000 L of MILK.
- Can you send the PO's to your supplier by email or is printing them the only option?

*If your system does not produce PO's, how do staff at receiving know what is expected to arrive?

Receiving Process

5. **Receive the raw materials you ordered**

- If your system has scanning capabilities that include traceability, can you scan the following GS1-128 labels to receive these raw ingredients into inventory in the MAIN warehouse? (Fig. 11.4)

- **TIP:** Walk around your warehouse looking at labels on raw materials. What percentage of your raw materials contain a GS1-128 barcode allowing a scanner to capture multiple pieces of information in a single scan? If they have GS1-128 labels, do they contain the important segments (01) GTIN-14 or (02) and Lot (10) or Serial Number (21) allowing a scanner to capture the item and lot information with a scanner? In the meat industry, GS1-128 labels are common, while many other segments of the food industry are just beginning to adopt these labels. Try to estimate for your business how much relabeling will be necessary to use scanners for both inventory control and traceability?
- If your system can't scan the above label to capture both the item code and lot information at once, what kinds of labels can it scan and what information can be captured?
- If your system can only scan GTIN-14 SCC codes, like the samples below, it is capturing the item information through that scan but not the lot number, so that scan process is not sufficient for traceability (Fig. 11.5). If these GTIN-14 are the only possible labels it can scan, how does the system manage

Fig. 11.4 GS1-128 barcodes

(01)01234567890128(3102)000100(10)L002

SALT

(01)45678901234560(3102)000300(10)L001

CULTURE

Fig. 11.5 GTIN-14 SCC barcodes

01234567890128

SALT

45678901234560

CULTURE

traceability? Does the lot information have to be keyed in (leaving room for error) and how does it get matched with these items?

- If you have to key the receiving information into a system, either because the system can't scan GS1-128 labels or because the ingredients arriving do not have a GS1-128 label, follow the system's process to receive the SALT and CULTURE into the MAIN warehouse location.
- <u>Receive:</u> 3 bags of CULTURE. As you can see from the GS1-128 label, L001 lot has an expiry date of September 1, 2015.
- <u>Receive:</u> 1 bag of SALT of lot L002. Its expiry date is also September 1, 2015.
- <u>Receive:</u> 1000 L of MILK of lot L003. Its expiry date is also September 1, 2015.
- Can the system accommodate these alpha-numeric lot numbers and how long is the field? If you can't receive an alpha-numeric lot number you will have to reassign the supplier lot number to an internal one—how will the system keep the two numbers cross linked in case your supplier issues a recall (they will issue it using their lot number)?
- If not scanning, do you have an option to enter the expiry date and does the system have expiry date reporting or features to automatically alert a user if they scan something past its expiry date?

Relabeling

- If the product did not have GS1-128 barcode labels, and the system supports scanning, can the system print you a GS1-128 label to apply to this product so that moving forward you can capture the item and lot information in a single scan? Does this GS1-128 label encode the important segments including item (01) or (02) and lot number (10) or serial number (21)? If it does not contain either the lot or serial number segment, what would the system use as a lot number?

- If a skid label is an option, what does the barcode on the skid label encode? Does it represent one unit on that skid or does it correspond in the system to all the items and lot numbers (the total number of all units) that are contained on the skid? Will the skid label be useful if you begin to case pick off that skid or is it only useful to move the entire skid around the warehouse until case picking begins?
- Does the system give you an option to enter Key Data Elements about the receiving as a whole (i.e. driver dropping off shipment, temperature of truck, condition of truck, etc.)?
- Does the system give you the option to enter KDEs about the individual lots received, beyond quantity, lot and expiry information (i.e. shelf-life, country of origin, Kosher certified, etc.)?

6. **Check that the raw materials and lots received are accurately in stock in the MAIN location. There should be some kind of lots in stock report to use for this. If your system also does costing, run an inventory valuation report. Are the materials that you ordered now in stock at the costs you would expect them to be?**

Item	Quantity	Unit	Cost	Total
MILK	1000	Liters	0.50	500.00
CULTURE	30	Kilograms	1.00	30.00
SALT	10	Kilograms	1.00	10.00

Storage Process

7. **Transfer MILK from the MAIN receiving location to the COOLER location.**

- Transfer the 1000 L of MILK from the MAIN warehouse to the COOLER warehouse.
- When attempting the transfer, do you need to enter the lot number or can you select it from a list of available lots? If keying in the lot number, try entering the lot number as LOO3 (switching the two zero's for letter O's).
- Does your system allow you to make this error? If so, the system is not verifying that the transfer transaction is possible. Meaning, you didn't receive LOO3, so you should not be able to move or use it. Complete this incorrect transaction if the program will allow it.
- If the system lets you save the mistaken transaction, review a lots in stock report, does it show, 1000 L of MILK in location MAIN lot L003 and 1000 L of MILK in location COOLER with LOO3? If so, your inventory has just doubled and this system will not help you with inventory or traceability. If it

moved the 1000 L of Milk but allowed you to reassign the lot number to LOO3, run a lot history to be sure that the two lots are in fact linked should you have to recall lot L003.

- If the system let you save that mistaken transaction, try now to edit the transfer to correct the lot number error. Does the system have editing capabilities and will it let you? If not, the only way to correct the above error is with a new transaction that moves L003 where you want it and another that involves a negative transaction (moving –5 units of LOO3). If your system allows these negative transactions, it could quickly create unclean and unusable traceability data.
- Try another transfer to transfer more inventory (10 units) than you have in stock. For example, try to transfer 1500 L of MILK back from COOLER to MAIN. Does the system let you? If so, it is not controlling the inventory and your data could quickly become corrupted. You did not have 1500 L of MILK to complete the transfer. Put back this MILK into the COOLER location if the transaction did go through.
- If the system did not let you make any of these mistaken transactions, transfer the 1000 L of L003 to the COOLER location.
- Run a lots in stock report to ensure that 1000 L of MILK is now in the COOLER location.

8. **Transfer SALT and CULTURE to rack/bin A1 within the warehouse MAIN. (if your system supports locations)**

- Run a lots in stock report to ensure that CULTURE and SALT are now in the MAIN warehouse, slot location A1

9. **After you transfer your product to another location, edit the original inventory receipt and say that you have received less than you transferred above. Does the system let you do it?**

- We originally received 1 bag (10 kg) of SALT and 3 bags (30 kg) of CULTURE and 1000 L of MILK and then transferred those amounts of MILK from MAIN to COOLER and SALT and CULTURE from MAIN to slot location A1.
- Now, attempt to edit the original receiving and say that we only received 950 L of MILK of lot L003.
- If the system lets you make this change, save the transaction and look at the lots in stock report afterward. Does it now show 1000 L of MILK L003 and – 50 L of MILK L003 in stock or has it changed the original entry to 950 L? Is the 950 L still in the COOLER location?
- If you cannot edit the receiving, try entering a new receiving for the –50 L of MILK using a new lot number, L004. Did the system let you save this transaction? If so, look again at the lots in stock report. If you had hundreds of receipts over the course of a week, how would you know that the L003 for 1000 L and L004 for –50 L were actually related transactions, where one transaction was attempting to fix the other?

- Try another transaction where you receive −2500 L of lot L003. If the system lets you say that you received a negative quantity higher than the original positive quantity, the system is not verifying what was originally entered, but rather allowing you to enter any value, positive or negative. This could lead quickly to corrupted traceability data.
- If costing is part of the system, run an inventory valuation report for warehouse COOLER to see if the edited receiving automatically updated. If it did not, can you edit the transfer to reflect the changed receiving?

Item	Quantity	Unit	Cost	Total
MILK	1000	Liters	0.50	500.00

Pick for Production Process

10. **Run a requirements report to ensure that you have all the raw materials necessary to complete a production run to fulfill the work order we scheduled in step 3.**
    - There should be no requirements for MILK, CULTURE or SALT as they have already been received.
11. **Many systems have First In, First Out lot suggestion capabilities to help with inventory rotation of raw materials. Others might have no such capabilities or have other options such as First to Expire. To understand the implications of any lot suggestion capabilities in your system, try this experiment.**
    - Enter a new receiving for MILK with lot L005 with expiry date June 30, 2015.
    - Run a pick list for picking raw materials for your production. Because MILK in L005 was received after the MILK in L003, if FIFO is operating, the system should suggest you pick or use lot L003 before suggesting or using lot L005. However, if First to Expire is operating, it will suggest you pick and use lot L005 first.

MILK

- Lot #: 100 of L005 in COOLER
- Lot #: 900 of L003 in COOLER

CULTURE

- Lot #: 30 of L001 in MAIN

SALT

– Lot #: 2 of L002 in MAIN

Manufacturing Process

12. **How is production recorded in the system? Is it entered at a work station/plant floor station or on a tablet or handheld computer? Does the process of recording finished goods made automatically deplete the raw materials that were used to make it?**

    - Record that you made ten packages of CHEESE using lot CHE-060315 (Fig. 11.6)
    - Do you have an option to print a label for the finished good? What kind of identifier can be on the label? A GS1-128 barcode including item, lot and date of production or expiry, a GTIN-14 (SCC), a GTIN-12 (UPC) or what combination of these?
    - Below is a GS1-128 which encodes the GTIN-14 for the item, the date of expiry of August 1, 2015 and lot number of CHE-060315 for this product:

    - If costing is part of the system, run an inventory valuation report to look at the cost of CHEESE to make sure it makes sense
    - Run a lots in stock report to show that the raw materials have been depleted and CHEESE is now in stock

Item	Lot	Stock
MILK	L003	100
SALT	L002	8
CHEESE	CHE-060315	10

(01)23456789012344(17)150801(10)CHE-060315

CHEESE

Fig. 11.6 GS1-128 Barcode encodes the GTIN-14 for the item, expiration date and lost number

13. **Move your system forward one day (to June 4, 2015), as you will package the cheese into cases on a different day than it is produced. (to check the system's abilities to manage production that extends across multiple days)**
14. **Run a requirements report to ensure that you have all of the CHEESE necessary to complete a packaging run to fulfill the sales order scheduled in step 1.**
    - CHEESE has already been made so there should be no entry on the requirements report for CHEESE, or an entry of zero.
15. **Pick the raw materials for this subsequent packaging process. Use the CHEESE barcode in step 13 as the interim product label.**
16. **Record the subsequent production of packaging of the CHEESE into CHEESECS:**
    - Record that you made 5 cases of CHEESECS using CHECS-060415
    - Do you have an option to print a label for the finished good? What kind of identifier can be on the label--a GS1-128 barcode including item, lot and date of production or expiry, a GTIN-14 (SCC), a GTIN-12 (UPC) or what combination of these?
    - Below is a GS1-128 which encodes the item GTIN 14 (in the (01) segment, the date of expiry as August 1, 2015 (in the (17) segment) and the lot number it for this product as CHECS-060415 (in the (10) segment. If your system has scanning to ship, this barcode will allow you to scan the product going out (Fig. 11.7).
    - If costing is part of the system, run an inventory valuation report to look at the cost of CHEESECS to make sure it makes sense.

Item	Quantity	Unit	Cost	Total
CHEESECS	5	Cases	106.40	532.00

- Run a lots in stock report to show that the stock of CHEESE has been depleted and CHEESECS is now in stock.

Item	Lot	Stock
CHEESECS	CHECS-060415	5

(01)34567890123457(17)150801(10)CHECS-060415

CHEESECS

Fig. 11.7 GS1-128 Barcode encodes the GTIN-14 for the item, expiration date and lot number as CHEESECS

Shipping/Invoicing Process

17. **Enter a shipment to ship the CHEESECS to the customer STORE that ordered it in step 1.**
    - If keying this entry, can you adapt the shipment from sales order in step 1, so that it suggests you ship the full amount?
    - Does the system suggest which lot to pick based on FIFO or First to Expire?
    - If scanning to ship, use the label provided in step 16 of this script.
    - If you scan the wrong product (i.e. scan a GS1-128 raw material barcode from step 5 instead of your finished good GS1-128 barcode), does the system alert you it is the wrong item or let you pick that other product?
    - Does the system produce shipping documents like a bill of lading with the lot information on it?
18. **Is the system EDI (Electronic Data Interchange) ready?**
    - If you sell to major retailers or hope to, is the system able to produce EDI files such as an ASN (Advanced Shipping Notice) and the corresponding MH10 serialized skid labels in the required format for each EDI partner you deal with?
19. **Enter an invoice for the shipment**
    - Can you adapt the invoice from the shipment above?
    - Can you email or print the invoice from this system? If it is sending information to another system for invoicing, is the information available immediately for invoicing or do you have to wait for a batch process to complete it?
    - Can the system produce EDI invoices, or will you need to rekey the invoice information into a web portal?
20. **Ensure that the finished good is no longer in stock.**
    - Check lots in stock report and if you shipped all of the finished good, then the finished good lots should no longer be in stock.
21. **Traceability Reporting Check**
    - Run a traceability recall report for raw material CULTURE, lot L001. Does it show that L001 was used in making the CHEESE in lot CHE-060315 on June 3, 2015, which was subsequently used in making CHEESECS in lot CHECS-060415 on June 4, 2015 and that 10 cases went out to our customer STORE on June 4, 2015?
    - Run a traceability recall report for finished good CHEESECS lot CHECS-060415 to ensure it shows that 0 units of that lot are still remaining in stock, but that 10 units were shipped to STORE. Is it easy to get the STORE contact phone and email info so you can expedite the recall?
    - Is there a traceability report that shows everything that was made on a particular day (possibly even on a particular line on a particular day)? This

report can be used if there is a problem with production such as a problem with a piece of equipment that contaminated all product going through that equipment. If you run that report for June 3, 2015, the date the CHEESE was made, does it show CHEESE having been produced but also alert you that there was subsequent production that took place the day following that also needs to be recalled as well (CHEESECS)?

You can adapt this script if you are selling catch weight products because in that situation, you will have a unique lot number or serial number for each case of each item, since a unique weight will be associated with that case.

Traceability information can be used for so many things beyond recall reporting, such as costing, EDI document preparation, and much more. The functionality of the tool you select to record this information and its ability to tie in to any other relevant systems will determine what benefits beyond traceability reporting are possible from the information being tracked. Take the time to understand the options and limitations to select the best system for your business.

References

1. Kirkness JL (2015) The traceability factor: increase profits through tracking, 3rd edn. Charlston
2. McEntire J, Bhatt T Pilot projects for improving product tracing along the food supply system [Internet]. Chicago: Institute of Food Technologists. 2012 [cited 2015 Apr 27]. Available from: http://www.fda.gov/downloads/Food/GuidanceRegulation/UCM341810.pdf. Accessed 13 Apr 2015

Chapter 12
Connecting the Dots with Whole Chain Traceability

Andrew W. Kennedy and Jennifer McEntire

Abstract The term, "Whole Chain Traceability" refers to a combination of internal traceability systems and external traceability systems applied in a consistent way across the food industry. This level of interoperability requires global, food industry-wide agreement and adoption of product and location identification standards; Critical Tracking Event (CTE) and Key Data Element (KDE) content, structure, and capture; and common data sharing communication technology frameworks.

Drivers for Whole Chain Traceability vary by stakeholder. Regulatory bodies charged with investigating foodborne illnesses see traceback data across the supply chain, epidemiological data and food and environmental testing data as three critical inputs in outbreak investigations. The food industry is finding, based on successfully implemented industry segments, discrete supply chains and pilot testing, that Whole Chain Traceability can increase profitability by improving the quality of food produced, supporting belief attributes, preventing counterfeiting and protecting entire markets. Food industry segments that stand the most to gain by implementing processes and systems that enable Whole Chain Traceability face significant hurdles due to their size, structure and competitive dynamics. Some of these supply chains include fresh produce, seafood, baby formula, flour, spices, coffee, cheese, chocolate, honey, olive oil, meat and large consumer packaged goods companies with extensive global supply chains. Consumers, who vote multiple times a day with their wallets, consider food functional, as medicine, fashion, entertainment and, often, as a political statement – an image that is often shattered by food scandals that could have been avoided or mitigated with Whole Chain Traceability.

Looking ahead, one can imagine that the cost and ubiquity of sensors, mobile devices, robots, artificial intelligence and Blockchain will accelerate the development of Whole Chain Traceability for the food industry globally. Currently, data sharing is managed explicitly between two or more parties. The challenge for the

A. W. Kennedy (✉)
Institute of Food Technologists, Washington, DC, USA
e-mail: akennedy@ift.org

J. McEntire
United Fresh Produce Association, Washington, DC, USA
e-mail: jmcentire@unitedfresh.org

J. McEntire, A. W. Kennedy (eds.), *Food Traceability*, Food Microbiology and Food Safety, https://doi.org/10.1007/978-3-030-10902-8_12

industry will be to agree on what traceability data is necessary to capture and store in the hyper-connected, immutable world of Blockchain – and how to retain best data stewardship and governance practices.

Keywords Whole chain traceability · Interoperability · Supply chain · Blockchain

Whole Chain Traceability

The term, "Whole Chain Traceability" refers to a combination of internal traceability systems and external traceability systems applied in a consistent way across the food industry. "Internal traceability is a requirement for food processors, packers, distributors, and others in the food supply chain. It involves tracking all inputs where those inputs were used in manufacturing (as relevant) and all finished goods made. Further, it requires that records be kept of where all finished products go. This is so you can trace forwards and backwards the history of all inputs and outputs through each facility, in the event of a food safety issue." (Chap. 11). Internal traceability can become complicated if transfers between locations within one corporate umbrella are considered internal; they essentially need to meet the same criteria expected of external traceability.

External traceability, in its most simple form, is called "one-up, one-down" and is legally required for most food processors and handlers. Each entity in the supply chain records information about the product movement into and out of a facility along with basic information about the product and any transformations that may have occurred at the facility. Combined with data sharing tools, globally unique product and location identifiers, this enables records from one facility to be linked to the records of another facility, thus enabling Whole Chain Traceability. Regardless of whether "internal" or "external," Whole Chain Traceability can only be achieved when adequate records of product movement are available that correspond to the physical movement of the product.

Whole Chain Traceability makes use of globally unique traceable object identifiers, Critical Tracking Events (CTE's) and Key Data Element (KDE's) to connect internal and external traceability systems across many disparate organizations. This level of interoperability requires global, food industry-wide agreement on and adoption of product and location identification standards; Critical Tracking Event and Key Data Element content, structure, capture; and data sharing communication technology frameworks.

The vast majority of the food industry consists of reputable operators who want to "do the right thing." This means producing and handling food in a safe manner and responding to any issues as expeditiously as possible to protect public health. These companies operate with integrity and endeavor to provide accurate information about food products: their origins, handling practices, ingredients, etc. In general, the organizations that compose our food supply chain share the goal of providing accurate information about products, so they can meet consumer and regulatory requirements. They also share a common goal of remaining profitable by looking for efficiencies in their systems and processes. The same systems that afford traceability also accomplish these goals.

However, food operators share a singular challenge to implementing traceability: no one has infinite time or money. Suppliers balance just in time production versus the costs of carrying inventory, which can result in "fill in" material from alternate sources so that an order can be shipped on time. Many are subject to weather conditions, market disruptions, technology disruptions and other external factors that can also trigger changes in supply changes and sources of products. With constantly changing consumer demands, it is hard to know what products or information will be required tomorrow. This rapid change in the food industry presents several industry-wide, uniformly challenging, obstacles to adopting and implementing Whole Chain Traceability.

Despite some similarities, different parts of the supply chain do have specific challenges, each of which needs to be overcome to enable traceability throughout the whole chain.

- Consumers look at all foods together at retail, and therefore interoperability is important. It is impractical to imagine that retailers will implement different systems and processes for different types of products. Nor should a consumer be expected to have multiple apps to help him or her uncover additional information about products.
- Retailers commingle bulk products on the display shelf – specifically loose, fresh produce. As long as those products do not bear a unique identifier that can be scanned at checkout, process changes in terms of stock rotation, inventory management, and segregation can be used to improve traceability.
- Distributors handle all types of products and face the same need for interoperability as retailers. Given that most products in distribution centers are "in and out", speed is critical and systems that slow down the movement of goods will not be adopted.
  - Dealing with changing orders, dropping off different shipments to different locations and needing to verify the correct product was shipped; complicating matters, pallets are not homogenous; they consist of a mixture of items that need to be tracked.
- Processors handle different products within manufacturing
  - Ingredients are used at different rates; continuous flow manufacturing operations need to be considered.
- Any product from farm to fork goes through a distribution web that intermingles with other products
  - "Fill in" product can be used if inventory or production is low, especially for commodity products.

The need for alignment across sectors has never been more critical given today's low cost and ubiquity of data sharing systems. Who defines what traceable events should look like? Who should capture them, when and in what format should they be shared? The Institute of Food Technologists traceability pilot project in 2012 for the United States Food and Drug Administration's Food Safety Modernization Act recommended establishing uniform Critical Tracking Event (CTE) and Key Data Element (KDE) requirements for all segments of the food industry [17].

The impact of Whole Chain Traceability on fresh foods and processed foods will be significant and challenging in different ways. Product identification, data capture and data sharing on a fishing vessel are not insignificant challenges to overcome. Neither is the challenge of traceability from farm to finished product for a food processor with ingredients from thousands of small farms spread across four continents.

The good news in the traceability world is that technology is getting better and cheaper at a rapid pace. Additionally, the increasing sophistication of food industry leadership and food safety driven quality management systems provide for better environment for adopting and supporting new technologies. Importantly, there is stronger public awareness and demand for food traceability and transparency than ever before – providing the motivation for the food industry to "Connect the Dots."

Whole Chain Traceability Is a Team Effort Requiring Advanced Technology

Individual companies will boast that they have terrific traceability and can trace products within a short time frame (often minutes to hours). The idea that Whole Chain Traceability requires full participation from *all* supply chain members is difficult to grasp. The saying "you are only as strong as your weakest link" rings very true when describing traceability within a supply chain. Thus, a system-based approach needs to be implemented, with common terminology and alignment around standards. Data sharing platforms such as blockchain can be used to relate information throughout the supply chain.

Testing a Whole Chain Traceability system is complicated. The food industry is accustomed to conducting mock recalls. These are generally limited to the raw materials and/or finished products under the control of one facility, or perhaps one company. Rarely do companies actively engage with suppliers or buyers to test their recall systems. Testing a whole chain system is substantially more complicated. Evaluating the ability to trace back products and determine if they have commonalities is an even greater challenge. However, acquainting the food industry with the differences between whole chain and internal traceability, and between a mock recall and a mock traceback, are outstanding needs.

A systems approach is required to trace products; including incorporating verification, authentication, reliability, robustness, precision, accuracy and granularity. The primary approach in place today is data focused- which relies on recordkeeping. A second approach relies on scientific authentication for verification using

DNA analysis, chemical fingerprinting, and other tests performed on the food itself. These are used to verify that records that accompany a product match its physical attributes.

Whole Chain Traceability Driver: Critical for Faster Food Safety Investigations

The primary benefits to an individual company or facility of improved recordkeeping, which lays the foundation for improved traceability, are largely economic, as summarized later in this chapter. Operational efficiencies can be gained, as discussed throughout several chapters. However, the benefits of Whole Chain Traceability are realized by consumers. Daily confidence in the transparency associated with their purchase, and the authenticity of ingredients and claims appeals to consumers. On thankfully rare occasions, Whole Chain Traceability is used to protect public health from foodborne disease.

Scientists are increasingly able to detect foodborne illness and foodborne outbreaks. This is not because these events occur with increased frequency; rather it is because the scientific tools have advanced to detect issues that were previously "under the radar." These include systems such as PulseNet which originally relied on Pulsed Field Gel Electrophoresis (PFGE), and increasingly utilizes a granular approach called Whole Genome Sequencing (WGS).

PulseNet is managed by the Centers for Disease and Prevention and accepts bacterial "fingerprints" from state and other laboratories. These fingerprints, which can be derived from patients experiencing foodborne illness, contaminated food samples, or environmental or facility isolates, are loaded into and compared within a central system. When the analysis was done via PFGE, a simple "yes/no" answer was provided indicating whether the fingerprints of the two isolates matched. With WGS, a genetic relatedness (close, distant, etc.) can be derived. Both approaches give investigators clues that an outbreak is occurring. The system identifies about 30 multi-state outbreaks each year, with around 1500 clusters of disease reported by state and local health agencies annually [3].

Outbreaks, particularly multi-state outbreaks, are investigated by asking ill patients what they ate. By comparing what they have in common—and how that differs from what non-ill people ate; FDA or the USDA Food Safety and Inspection Service investigators can study suspect foods. Early in an investigation, lot numbers are unknown and little information exists. Information is gathered by collecting records, beginning with those from point of sale or point of service. Currently, investigators must move through the supply chain, slowly gathering records at each step (assuming they are available); and painstakingly sort through them to establish linkages and identify and resolve gaps and discrepancies. In most cases, the product is no longer available; the only thing that remains are the electronic or paper records that reveal the products pathway. By tracing several different 'legs' of the supply chain (locations where several ill people shopped or ate); investigators hope to find a point where the supply chains converge to a common facility, farm, or other entity. The number of points in a supply chain, and the flows of product, may be numerous and complicated. One vision for Whole Chain Traceability, from regulators, will be to eliminate the need for this time-consuming approach—time during which people may continue to be exposed to contaminated product and become ill.

In the 2018 outbreak associated with Romaine lettuce from the Yuma, Arizona growing region, one of the many complications of the investigation was that supply chains did not converge. Multiple growing locations were identified during the investigation [6]. Several discrete supply chains were identified, but they did not share any common links. Improved traceability could have aided the investigation in a few ways. First, better processes could have reduced the variables that FDA discovered during the traceback, stemming from the uncertainty about whether a particular case of product originated from supplier A or supplier B, for example. Each time this kind of uncertainty exists, it expands the number of pathways that need to be traced, wasting time and resources, and preventing regulators from focusing on those entities that definitely handled implicated product. Additionally, Whole Chain Traceability could have illuminated supply chains more rapidly than the current "one-up, one-down" approach. During the outbreak investigation, the FDA challenged the industry to leverage technology, and asked questions about the capabilities of blockchain. Although laws including the Food Safety Modernization Act prevent FDA from prescribing a particular traceability system or technology for use by the food industry, it is evident that ignoring the availability of such tools exacerbates the impact of outbreaks, both in terms of public health, consumer confidence, and perception by regulators.

Currently, remarkably few outbreak investigations are solved. This is because investigators rarely find a point of convergence. This is problematic for several reasons. First, product may remain in the distribution system and continue to put lives at risk (the chapter on produce traceability notes the infamous *Salmonella Saintpaul* outbreak). Taking a longer view, the inability to find a point of convergence means that a root cause analysis into the issue cannot be conducted. The reason for the problem remains unknown, and preventive measures are not enacted to prevent a recurrence. Therefore, Whole Chain Traceability not only addresses an urgent public health need, but also potentially improves the safety of the food supply in the future.

Whole Chain Traceability Driver: Food Industry Profitability

Benefits of traceability in the food supply chain are many, including food safety, supply chain efficiency, decreased food waste and consumer transparency. However, one of the important drivers for increased traceability is increased profitability. One example of a segment of the food industry that has rigorously applied Whole Chain Traceability and measured the cost benefit is the cattle industry [11].

Lower cost DNA analysis of breeder cattle, beef tenderness and quality measurement technology and RFID-based national livestock traceability programs have enabled researchers an opportunity to quantitatively evaluate the financial benefits of traceability. One such study is "The Value of a Whole-Chain Traceability System in Transmitting Genetic Information About Beef Tenderness," conducted in 2009 by Candi Ge, originally of Xi'an University of Technology, Xi'an, China and submitted to Oklahoma State University in 2014 as part of a Master of Science program. Ge's study evaluated the financial profit (or loss) of implementing a Whole Chain Traceability system in a vertically integrated three-stage supply chain, in coordination with artificial insemination and tenderness testing. Results showed that the extra value obtained would be $58.53/head or approximately 5%. This analysis assumes a retail preference and price premium for tender beef and a successful artificial insemination program – in addition to the Whole Chain Traceability program [11].

Managing consumer belief attributes are as challenging and financially rewarding as beef tenderness. For example, increased diligence by Muslims in observing dietary obligations has led to rapidly growing demand for food produced in accordance with Halal standards. Malaysia is a major producer and consumer of Halal foods. Raids carried out by Malaysian authorities have uncovered "Halal certified" food manufacturers producing foods that do not comply with Halal regulations. The Halal industry involves an extensive farm-to-table value chain. The lucrative global Halal market requires an effective electronic traceability system in Malaysia to mitigate the risks associated with assuring Halal compliance [22].

Increasing brand value often leads to counterfeiting, including products in the food industry – the fourth most valuable market for counterfeiters according to PMMI's 2016 Brand Protection and Product Traceability Market Research Report. Olive oil, honey, fish, vinegar, vanilla and coffee top the list. The international market for counterfeit goods ranges between $461 billion and $1.7 trillion according to the Organization for Economic Co-Operation and Development (OECD), representing 85% growth since 2008. Traceability technologies developed to counteract rampant food fraud including sophisticated identification tools including barcodes, holograms, watermarks, embossing, radio-frequency identification ("RFID"), Near-field communication ("NFC") 11Whole Chain Traceability systems. Traceability system features that assist with fraud prevention include fully integrated infrastructure, product movement recording and retrieval at each step in the supply chain and immediate data retrieval during a recall [7].

According to the international non-governmental organization, Oceana, the complexity and opacity of the seafood supply chain "opens the door to illegal and irresponsible fishing practices, seafood fraud, public health risks, and even slave labor and organized crime. These problems threaten the oceans and consumers' wallets and undermine honest fishermen and businesses that play by the rules." A significant part of the solution to protecting the seafood industry is traceability and increased transparency. Information sharing at all points in the supply chain reduces the risk of seafood fraud, mislabeling and helps prevent the sale and purchase of illegal products [8].

Food Industry segments that stand the most to gain by implementing Whole Chain Traceability are facing significant hurdles due to their size, structure and competitive dynamics. Some of these supply chains include:

Fresh produce – despite 10 years of produce industry effort dedicated to the North American Produce Traceability Initiative and 7 years of the FDA promulgating rules under the Food Safety Modernization Act (FSMA), the investigation into the 2018 *E. coli* O157:H7 outbreak tied to romaine lettuce was slowed due to lingering gaps and inconsistency in traceability records, according to the FDA [6].

Seafood – an Oceana seafood fraud investigation published in 2013 found that more than one third of 1200 seafood samples taken from 674 retail outlets in 21 States were mislabeled. Snapper and tuna had the highest mislabeling rates (87 and 59%). Often farmed tilapia is sold as red snapper and escolar sold as white tuna [24].

Baby formula – a Lactalis, a French company, recalled 7000 tons of baby formula potentially tainted with *Salmonella* which sickened at least 38 children in 2018 [1].

Flour – According to the U.S. Center for Disease Control, in 2016, a multijurisdictional team investigated an outbreak of Shiga toxin–producing *Escherichia coli* (STEC) serogroup O121 and O26 infections linked to contaminated flour from a large U.S. producer [5].

Spices – According to survey work performed by the FDA Center for Food Safety and Applied Nutrition, between 1.7% and 18% of imported spices are estimated to have detectable levels of *Salmonella*. For this reason, the U.S. spice industry has invested in pathogen reduction technology, resulting in dramatically lower *Salmonella* prevalence in spices sold at retail [26].

Coffee – With the increased popularity of high-quality coffee, counterfeiters look to extend the amount produced with inexpensive, low quality fillers such as maize, soybeans, sugar and acai seeds. Using analytical chemistry, firms are now able to detect tampering with 95% accuracy or better. However, the risk to the coffee industry is significant if coffee buyers and sellers are not constantly diligent about the product they buy [23].

Cheese – Leading grocery chains are inadvertently selling products labeled as 100% grated parmesan cheese that have been found to include no parmesan cheese at all. The FDA's report on a Pennsylvania cheese factory in 2012 determined that in some cases it was labeling products as 100% grated Parmesan Cheese – when, in fact, no parmesan cheese was found in the product. Instead, the product consisted of a mixture of Swiss, mozzarella, white cheddar and cellulose, a common anti-clumping agent made from wood pulp. According to an owner of one of the leading suppliers of hard Italian cheeses, Neal Schuman, 20% of U.S. production ($375 million in sales) is mislabeled [2].

Chocolate – Since 2012, Nestlé's Child Labor Monitoring Remediation System has uncovered 7000 children in its cocoa supply chain in Côte d'Ivoire and Ghana. Despite this effort, multiple lawsuits have been filed claiming that Nestlé has not fully disclosed the extent and nature of child labor in the supply chain that produces popular candy products including *KitKat*, *Crunch* and *Butterfinger* [18].

Honey – Honey is one of the most frequently counterfeited foods in the world due to its similarity to other liquid sweeteners and the labor-intensive process used to produce it. According to *Businessweek*, this led to the largest single incident of food fraud in U.S. history where over $80 million of banned Chinese honey was transshipped through Indonesia, Malaysia and India by a German distributor across a seven-year period [9].

Olive Oil – Research from the University of Nebraska-Lincoln demonstrated the devastating impact that a small group of bad actors can have on an entire industry – specifically, olive oil. The financial impact on all olive oil producers, regardless of country of origin, was significant when consumers were informed of mislabeling and adulteration of Italian olive oil. The most dramatic impact, a 50% decline in price, was experienced by Italian producers in the experiment, but U.S. producers also experienced a 9% decline in price – demonstrating the spillover effect [13].

Meat – In 2013, a large-scale scandal erupted when several companies in the United Kingdom and across Europe intentionally mislabeled horsemeat as beef. Consumers with cultural aversion to eating horsemeat were outraged. The brand impact to EU and UK retailers and branded food companies was significant. According to a Food Standards Agency spokeswoman, "This case has highlighted the importance of food businesses abiding by food traceability rules which are there to protect both the consumer and the business" [20].

Consumer Products – In response to consumer demand for organic, allergen-free and ethically sourced foods, large consumer product companies are investing in supply-chain food traceability, transparency and certification systems. "Sixty-five per cent of Americans are interested in the history of their food, According to Packaged Facts' report 'Nutritional Labeling and Clean Label Trends.' With little indication of this number decreasing, food manufacturers should begin taking steps to track the ingredients they are using and report any topical data to consumers" [25].

Whole Chain Traceability Driver: Consumer Confidence

Growing consumer concern about obesity and the impact of diet on health has led to a rise in demand for functional foods. Food manufacturers initially responded with low-fat spread butter, enriched milk, probiotic yogurts, juice with added calcium, fortified cereals with fiber and minerals, bars with added fiber, Xylitol sweets and chewing gum, and energy drinks [15]. Recently, however, the food industry in partnership with physicians and medical institutions is moving in the direction of offering food as medicine. Programs such as Ralph's Supermarket, a retailer based in Huntington Beach, California, "Shop with Your Doc," — incorporates in-person nutritional advice from a physician from nearby Mary and Dick Allen Diabetes

Center at St. Joseph Hoag Health. According to the physician, people can help reverse diabetes, hypertension and prevent cancer by choosing less processed, more plant-based food [12].

The interesting parallel to food as medicine is food as fashion. Due to the high availability and variety of food available in developed countries, consumers have the luxury of seeing food as a form of self-expression, or fashion. Often these food trends or novelties are framed by marketers and consumers alike as "healthy" choices with little or no concrete evidence to support the assertion. As we write this chapter, carbs and fat, genetically engineered foods, sugar, margarine, fried and fast foods are bad and protein of virtually any sort, almonds, oatmeal, avocados, blueberries, kale, eggs, chicken and fish are all good. The industry is conflicted on chocolate and red wine because they are high in healthy antioxidants but contain sugar and alcohol. Eating blueberries, strawberries, grapes, nuts, dark green veggies, salmon and green tea may, in fact, be better alternatives to alcohol and chocolate. Ultimately, consumers can choose what they want to eat regardless of season, time of day or location because of the efficiency of our modern food supply opening the possibility of food fads and trends [10].

A relatively recent societal influence on the foods we choose to eat and how we prepare them is the celebrity chef. Prior to TV and the internet, people learned to cook their family and friends. Now, previously unknown chefs educate and entertain audiences with creative preparations, exotic dishes and appealing personalities. One can debate how much the audience absorbs new cooking skills, but these shows certainly have had an impact on food choices and are generally popular and entertaining [16].

For those who prefer to think about food in the larger societal context, food has become increasing political and, in many ways, reflects the polarization of society in general. Wealthy countries are experiencing a major obesity crisis while developing countries still struggle to meet basic nutritional needs. Rural farmers are under pressure to reduce their impact on the environment and deliver "healthier," "sustainable" foods to wealthy, urban consumers. Add the genetic modification of crops, the impact of a changing climate on agriculture, trade barriers, worker health, safety and availability, animal welfare, declining stocks of certain fish species, and several other concerns add even more complexity to food politics and create a need for a more traceable and transparent food supply [19].

Looking Ahead

Looking ahead, one can imagine that the cost and ubiquity of Industry 4.0 technologies including Blockchain and distributed applications; wide area communication networks and connected sensors; digital image capture and voice recognition; mobile devices and applications; artificial intelligence and predictive analytics; automation and robotics will accelerate the development of Whole Chain Traceability for the global food industry. Despite the rising complexity of today's food supply chain, the pace of technology is accelerating. For example, in the space of 25 years the world saw the development of the internet, social media, mobile devices, self-driving cars, smart speakers, inexpensive whole genome analysis, and ubiquitous genetic modification via CRISPR. Meanwhile the food industry is still heavily reliant on paper-based systems and humans [4, 14].

Food traceability relies on data sharing. Currently, data sharing is managed explicitly between two or more parties using Electronic Mail, File Transfer Protocol sites, Electronic Data Interchange (EDI), Business Intelligence (BI) Repositories & Extract Transform Load (ETL) tools, Industry Portals or direct system-to-system integrations via Application Programming Interfaces (API). While these tools have provided a good framework for commerce, the challenge for the industry will be to agree on what traceability data is necessary to capture and store in the hyper-connected, immutable world of Blockchain – and how to retain best data stewardship and governance practices [21].

References

1. Alderman L (2018) 'My baby almost died': formula scandal sends shudders through France. New York Times [online]. Available at: https://www.nytimes.com/2018/02/01/business/france-baby-formula-lactalis.html. Accessed 2 Aug 2018
2. Bloomberg.com (2018) The parmesan you sprinkle on your cheese could be wood. [online] Available at: https://www.bloomberg.com/news/articles/2016-02-16/the-parmesan-cheese-you-sprinkle-on-your-penne-could-be-wood. Accessed 2 Aug 2018
3. Cdc.gov (2018) PulseNet | PulseNet | CDC. [online] Available at: https://www.cdc.gov/pulsenet/index.html. Accessed 30 July 2018
4. Chester R (2017) The future of food safety: the revolution is on our doorsteps. [online] New Food Magazine. Available at: https://www.newfoodmagazine.com/article/41390/future-food-safety-revolution-doorsteps/. Accessed 8 Aug 2018
5. Crowe S, Bottichio L, Shade L, Whitney B, Corral N, Melius B, Arends K, Donovan D, Stone J, Allen K, Rosner J, Beal J, Whitlock L, Blackstock A, Wetherington J, Newberry L, Schroeder M, Wagner D, Trees E, Viazis S, Wise M, Neil K (2017) Shiga toxin–producing *E. coli* infections associated with flour. N Engl J Med 377(21):2036–2043
6. Eskin S (2018) FDA food industry must act to improve product traceability. [online] Pew Charitable Trust. Available at: http://www.pewtrusts.org/en/research-and-analysis/articles/2018/07/12/fda-food-industry-must-act-to-improve-product-traceability. Accessed 30 July 2018
7. Ferrante M (2018) Combatting the growth of fake food. Food engineering. [online] Available at: https://www.foodengineeringmag.com/articles/97107-combatting-the-growth-of-fake-food. Accessed 30 July 2018
8. Fish stories: Success and Value in Seafood Traceability (2016) [online] Oceana. Available at: https://usa.oceana.org/publications/reports/fish-stories-success-and-value-seafood-traceability. Accessed 30 July 2018
9. Forbes.com (2016) Exclusive book excerpt: honey is world's third most faked food. [online] Available at: https://www.forbes.com/sites/larryolmsted/2016/07/15/exclusive-book-excerpt-honey-is-worlds-third-most-faked-food/#5cc5f28f4f09. Accessed 2 Aug 2018
10. Fox R (n.d.) Food and eating: an anthropological perspective. [online] Sirc.org. Available at: http://www.sirc.org/publik/food_and_eating_6.html. Accessed 8 Aug 2018

11. Ge C (2014) The value of a whole-chain traceability system in transmitting genetic information about beef tenderness. Masters. University Oklahoma
12. Gorn D (2017) NPR Choice page. [online] Npr.org. Available at: https://www.npr.org/sections/thesalt/2017/01/17/509520895/food-as-medicine-it-s-not-just-a-fringe-idea-anymore. Accessed 8 Aug 2018
13. Imran Ali Meerza S, Gustafson C (2018) Consumer response to fraudulent producer behavior in the agri-food marketing system | Agricultural Economics. [online] Agecon.unl.edu. Available at: https://agecon.unl.edu/cornhusker-economics/2018/consumer-response-fraudulent-agri-food-producer-behavior. Accessed 25 Aug 2018
14. Knott M (2018) How Industry 4.0 offers a smarter future. [online] foodmanufacture.co.uk. Available at: https://www.foodmanufacture.co.uk/Article/2018/07/26/Food-firms-begin-to-embrace-the-power-of-Industry-4.0. Accessed 8 Aug 2018
15. Küster-Boluda I, Vidal-Capilla I (2017) Consumer attitudes in the election of functional foods. Span J Mark – ESIC 21:65–79
16. Lengyel A (n.d.) The impact of celebrity chefs on domestic food habits – UWL Repository. [online] Repository.uwl.ac.uk. Available at: https://repository.uwl.ac.uk/id/eprint/4991/. Accessed 8 Aug 2018
17. McEntire J, Bhatt T (2012) Pilot projects for improving product tracing along the food supply system [Internet]. Chicago: Institute of Food Technologists. [cited 2015 Apr 27]. Available from: http://www.fda.gov/downloads/Food/GuidanceRegulation/UCM341810.pdf. Accessed 13 Apr 2015
18. Nieburg O (2018) Nestlé sued again for allegedly 'using child and slave labor to make chocolate'. [online] confectionerynews.com. Available at: https://www.confectionerynews.com/Article/2018/02/13/Nestle-sued-again-over-child-labor-in-cocoa-supply-chain#. Accessed 8 Aug 2018
19. Paarlberg R (2013) Food politics, 2nd edn. Oxford University Press, Oxford
20. Robinson N (2015) First person to face jail over horsemeat scandal. [online] foodmanufacture.co.uk. Available at: https://www.foodmanufacture.co.uk/Article/2015/01/30/First-horsemeat-conviction. Accessed 25 Aug 2018
21. Russom P (2011) Ten Ways Data Integration Provides Business Value | Transforming Data with Intelligence. [online] Transforming Data with Intelligence. Available at: https://tdwi.org/Articles/2011/05/18/Ten-Ways-Data-Integration-Provides-Business-Value.aspx. Accessed 8 Aug 2018
22. Suhaiza Z, Affrifin Z, Abd Wahid N, Othman R, Fernando Y (2010) Halal traceability and halal tracking systems in strengthening halal food supply chain for food industry in Malaysia. J Food Technol 8(3):74–81
23. Stoye E (2018) Catching criminals' coffee adulteration. [online] Chemistry World. Available at: https://www.chemistryworld.com/research/catching-criminals-coffee-adulteration/7640.article. Accessed 2 Aug 2018
24. Warner K, Timme W, Lowell B, Hirshfield M (2013) Oceana study reveals seafood fraud nationwide. [online] Available at: http://oceana.org/sites/default/files/National_Seafood_Fraud_Testing_Results_FINAL.pdf. Accessed 30 Jul 2018
25. Wiber A (2018) Ingredient traceability: Having all the answers. [online] Food business news. Available at: https://www.foodbusinessnews.net/articles/11428-ingredient-traceability-having-all-the-answers. Accessed 8 Aug 2018
26. Zhang G, Hu L, Pouillot R, Tatavarthy A, Doren J, Kleinmeier D, Ziobro G, Melka D, Wang H, Brown E, Strain E, Bunning V, Musser S, Hammack T (2017) Prevalence of salmonella in 11 spices offered for sale from retail establishments and in imported shipments offered for entry to the United States. J Food Prot 80(11):1791–1805

Index

J. McEntire, A. W. Kennedy (eds.), *Food Traceability*, Food Microbiology and Food Safety, https://doi.org/10.1007/978-3-030-10902-8

CPSIA information can be obtained
at www.ICGtesting.com
Printed in the USA
LVHW062258080719
623399LV00005BA/220/P

9 783030 109004